Think
like
EINSTEIN

Think
like
EINSTEIN

Step into the mind of a genius

Robert Snedden

This edition published in 2020 by Arcturus Publishing Limited
26/27 Bickels Yard, 151–153 Bermondsey Street,
London SE1 3HA

AD007329UK

Printed in the UK

CONTENTS

Who was Albert Einstein?

'Why is it that nobody understands me, and everybody likes me?'

Albert Einstein – from an interview published in the *New York Times*, 12 March 1944

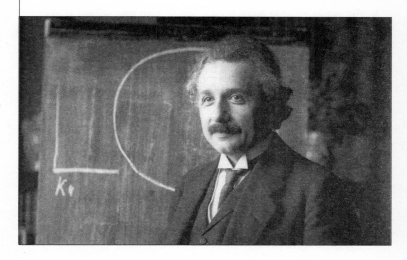

Albert Einstein was everyone's idea of what a scientist should be: absent-mindedly puffing on a pipe as he contemplated matters which seemed far beyond the comprehension of ordinary mortals. He was a genius and, beyond doubt, one of the greatest scientists who ever lived. But he had a very human side, too. It is said that in his later years, when he was living in Princeton, New Jersey, children would rush to see him and he would entertain them by wiggling his ears.

Early years

Einstein was born on 14 March 1879 in Ulm, Germany. He was the first child of Hermann and Pauline Einstein, a Jewish couple. Young Albert was supposedly slow in learning to talk, sometimes to the irritation of his family. 'I very rarely think in words,' he later said. 'A thought comes in and I may try to express it in words afterwards.' In June 1880, the family moved to Munich, where Hermann and his brother Jakob founded an electrical engineering company. Einstein's sister, Maja, was born in November 1881. On seeing her for the first time, Einstein exclaimed, 'Yes, but where are the wheels?'

At the age of around four or five, while ill in bed, Albert was given a compass to play with by his father. He was fascinated by it and the mysterious invisible forces that made the needle swing round. He later said this had made a deep and lasting impression on him and awakened his curiosity in the world.

Albert liked to work on puzzles and make complex structures with his building set. In 1885, his mother, a skilled pianist, arranged for Albert to take violin lessons. It began a love of music that lasted throughout his life and soon mother and child were playing Mozart duets together. He started his primary education at a Catholic school in Munich that same year. He was often top of his class. There is a persistently repeated story that Einstein was a poor student

Albert with his younger sister, Maja, around 1886

of mathematics in school, but it isn't true. When this was reported to Einstein in 1935 he laughed and declared he had never failed at maths, but had usually ranked first or second in his class. 'Before I was 15 I had mastered differential and integral calculus.'

In June 1894, the family moved to Italy, leaving 16-year-old Albert in Munich to finish school. Einstein missed his family and grew depressed. He obtained a certificate from his family doctor citing nervous disorders and was released from school. In spring 1895, he travelled to join his family.

Swiss citizenship

He took his entrance exam for the polytechnic in Zurich in October 1895. Although he did well in maths and science, he didn't gain admission, so attended the Kantonsschule in the town of Aarau to study for the qualification he needed for the polytechnic. In January 1896, he renounced his German citizenship and that autumn, having passed his exams, registered as a resident of Zurich. He became a student at the polytechnic, with the aim of becoming a teacher in maths and physics. He then formally applied for Swiss citizenship, which was granted on 21 February 1901.

After gaining his diploma degree in 1900, Einstein began to look for work, applying, without success, for assistantships at the polytechnic and other universities. Finally, in May 1901, he found a temporary job as a substitute teacher for two months at a high school in Winterthur. This was followed by another temporary position at a private school in Schaffhausen.

There he wrote his doctoral dissertation on the kinetic theory of gases, but it wasn't accepted. In 1902, Einstein moved to the Swiss capital, Bern, in the hope of finding a job at the patent office. In the meantime he gave private lessons in maths and physics.

In January 1902, Einstein had a daughter, Lieserl, with Mileva Maric, who had been his fellow student at the polytechnic in Zurich. The existence of Einstein's illegitimate child only came to light when private letters mentioning her were published in 1986. Einstein apparently told no one about the child and, it seems, never laid eyes on her (she was born at Maric's family home in Hungary).

Marriage to Mileva

On 16 June 1902, Einstein found a job as technical expert third class at the Bern patent office, on a trial basis only. At the end of 1902, his father became gravely ill in Milan; Einstein travelled from Bern to Milan to be with him. On his death bed, Einstein's father finally consented to his son's marriage to Mileva; on 6 January 1903 Einstein married her, much to the disapproval of both families. In May 1904, his first son, Hans Albert, was born, followed in July 1910 by his second son, Eduard.

Einstein enjoyed his job at the patent office. He took his work seriously but still managed to find enough time and energy to continue with his physics research. Writing to his friend Michele Besso years later, he reminisced about: 'these days in that temporal monastery, where I hatched my most beautiful ideas and where we spent such pleasant time together.'

In April 1905, Einstein submitted his doctoral thesis, 'A New Determination of Molecular Dimensions', to the university in Zurich. It was accepted in July and marked the beginning of a remarkable outpouring of ideas. No one before or since has changed science in such a profound way and in as short a time as Albert Einstein did in 1905.

1905 – Miracle year

The list of Einstein's achievements in his 'miracle year' makes for impressive reading:

1. **'On A Heuristic Point of View Concerning the Production and Transformation of Light'**, completed 17 March.

   This paper on light quanta and the photoelectric effect eventually won him the Nobel Prize in physics, and predates his PhD thesis.

2. **'A New Determination of Molecular Dimensions'**, completed 30 April.

   His doctoral thesis became the paper that was most often quoted in modern scientific literature.

3. **'On the Motion Required by the Molecular Kinetic Theory of Heat of Particles Suspended in Fluids at Rest'**, submitted 11 May.

   Einstein's 'Brownian motion' paper followed on from his thesis work.

4. **'On the Electrodynamics of Moving Bodies'**, submitted 30 June.

   The first paper on special relativity.

5. **'Does the Inertia of a Body Depend upon its Energy-Content?'** submitted 27 September.

   The second paper on special relativity, containing the famous $E = mc^2$ equation.

6. **'On the Theory of Brownian Motion'**, submitted 19 December.

   A follow-up to his earlier paper on 'Brownian motion'.

In April 1906, Einstein was promoted to technical expert second class at the patent office. His first application, in 1907, for a professorial position at the University of Bern was turned down. In early 1908, however, he was successful and gave his first lecture at the end of that year. Deciding that he wanted to devote his life to science, he resigned his position at the patent office in October 1909 and began work as an adjunct professor of theoretical physics at the University of Zurich. He was offered a chair at the German university in Prague in 1911, which he accepted, but he returned to Switzerland after a year to take up a professorial position at the Zurich polytechnic.

Impressed by what Einstein had achieved, the physicist Max Planck (1858–1947) offered him a professorial position without teaching responsibilities at Berlin university; this would make him a member of the Prussian Academy of Sciences and head of the planned Kaiser-Wilhelm-Institute of Physics. It was too enticing an offer to pass up and Einstein accepted enthusiastically, taking his family to Berlin in April 1914.

Unfortunately, his marriage did not fare as well as his career. In July 1914, after only a few months in Berlin, Mileva returned to Zurich, taking the children with her. The couple eventually divorced in February 1919. From 1917 until 1920, Einstein suffered ill health and was nursed by his cousin Elsa Loewenthal, who he married on 2 June 1919. Elsa had two daughters, Ilse and Margot, from her first marriage. Charlie Chaplin, who met Elsa in 1931, described her as 'a square-framed woman with abundant vitality; she frankly enjoyed being the wife of the great man and made no attempt to hide the fact; her enthusiasm was endearing.'

Between 1909 and 1916, Einstein was hard at work on the General Theory of Relativity, which he eventually published in March 1916 as 'The Foundation of the General Theory of Relativity'. One consequence was the theory's prediction that the light from a distant star would be bent by the gravity field of a massive body, such as the Sun. This was confirmed in

1919 by the British scientist Arthur Eddington, who observed Einstein's predicted bending of starlight near the Sun during a total eclipse (see pages 116–17). J. J. Thomson, president of the Royal Society, declared it 'the most important result related to the theory of gravitation since the days of Newton. . . [it] is among the greatest achievements of human thinking.'

In the early days of World War I, Einstein spoke publicly in support of pacifism, one of his lifelong concerns. He met with a hostile response; the chief of staff of the Berlin military district advocated removing pacifists from the streets. But the General Theory of Relativity had placed Einstein on the public stage – invitations and honours flooded in from all over the world. Nevertheless, he and his theory were also subjected to anti-Semitic abuse; even some German Nobel laureates were hostile towards him and demanded a 'German physics'.

Einstein was greatly affected by the unrest in Germany in the early 1920s. In 1922, he and Elsa departed on a five-month trip abroad. 'I very much welcomed the opportunity of a long absence from Germany,' he said, 'which took me away from temporarily increased danger.' On this trip he received word that he had been awarded the Nobel Prize.

From 1920 onwards, Einstein was attempting to formulate a unified field theory, one that would unite gravity with electrodynamics. It was a quest that would occupy him until his death, and one he never succeeded in fulfilling. At this time, Niels Bohr, Louis de Broglie, Werner Heisenberg, Wolfgang Pauli and other physicists were setting the foundations of the new physics of quantum mechanics. Einstein was unable to accept the theories of quantum mechanics and was constantly challenging it. Today, the tenets of quantum mechanics are as widely accepted as are Einstein's own theories, though science is still vexed as to how the two views can be reconciled.

Einstein in New York

In December 1932, Einstein and Elsa went on a lecture tour of the United States. In the 1932 elections in Germany, the Nazis became the strongest political party and in January 1933 Hitler seized power. Einstein would never return to Germany. In May 1935, he and Elsa sailed to Bermuda; it would be Einstein's last trip outside the United States. Not long afterwards, Elsa became ill and died of heart disease on 20 December 1936.

In 1939, Einstein's sister Maja, who had been living in Italy, was forced to flee to escape Mussolini's fascists and came to live with her brother in Princeton, New Jersey. With Europe on the brink of war, Einstein was convinced that scientists in Germany would be working on the technology for an atomic bomb. On 2 August 1939, he signed a letter to US President

Franklin D. Roosevelt, drawing his attention to the danger and urging the United States to embark on developing its own atomic bomb programme.

On 1 October 1940, Einstein became an American citizen, though he also retained his Swiss citizenship. In October 1946, he wrote an open letter to the general assembly of the United Nations, urging the formation of a world government; this enduring belief of his was something he saw as the only way to ensure lasting peace.

In August 1948, Einstein's first wife, Mileva Maric, died in Zurich. In December of that year, Einstein had to undergo abdominal surgery. In March 1950, he signed his will making his secretary, Helen Dukas, and Dr Otto Nathan his joint executors. Helen Dukas had been Einstein's secretary since April 1928 and, following Elsa's death, became his housekeeper. She remained with Einstein until his death.

Dukas devoted herself to the task of sorting and cataloguing Einstein's papers. Largely thanks to her work, these documents can now be found in the Albert Einstein Archive of the Hebrew University in Jerusalem. In November 1952, Einstein was offered the presidency of Israel, which he declined.

On 11 April 1955, he wrote to the philosopher Bertrand Russell saying he agreed to be a signatory to a manifesto urging all nations to renounce nuclear weapons. That same week, in an unfinished manuscript, Einstein set down the last phrase he would ever write: 'Political passions, aroused everywhere, demand their victims.' On 15 April, he was transported to hospital in severe pain. He died a few days later, on 18 April, aged 76. The diagnosis was a ruptured aneurysm of his abdominal aorta.

In accordance with his wishes, his remains were cremated that same day and the ashes scattered two weeks later at an undisclosed location.

To the end of his life, Einstein retained his sense of curiosity and wonder at the universe around him. He wrote to a friend: 'People like you and me never grow old. We never cease to stand like curious children before the great mystery into which we were born.'

CHAPTER 1

Does the universe run like clockwork?

Since ancient times, people have puzzled over the workings of the universe.

The music of the spheres

The great Ancient Greek thinker Plato, born around 427BC, declared that the heavens were perfect and the stars and planets moved in 'perfect curves on perfect solids'. He believed that these spheres produced music as they turned – an idea that would persist for many centuries to come.

However, the celestial spheres simply didn't fit the evidence of the skywatchers' eyes. A handful of stars behaved oddly, appearing to move against the background of the other 'fixed' stars, sometimes even looping backwards in their paths before continuing on their course. These peculiar stars were called 'asteres planetai' by the Greeks, meaning 'wandering stars'. We know them as planets.

The Ancient Greeks concocted all sorts of complex schemes to explain planetary motion involving spheres moving within spheres within yet more spheres, all rotating in slightly different directions. Around AD100, the astronomer Ptolemy set out a map showing an Earth-centred universe of nested spheres – an idea that would remain largely unchallenged for 1,400 years. It lasted so long because it seemed to work. Ptolemy's system gave accurate predictions of where the planets would be found at any given time.

Celestial revolutions

In 1543, astronomy was awakened from its Ptolemaic slumbers by the arrival of a remarkable book by Nicolaus Copernicus called *De revolutionibus orbium coelestium* (*On the Revolutions of the Celestial Spheres*). In 1507, the Polish astronomer and mathematician Copernicus had pretty much the same idea as Aristarchus 1,800 years previously: if the Sun was at the centre of the universe and the Earth and planets orbited around it, some of the puzzles of planetary motion could be explained. Mars, Jupiter and Saturn were further away from the Sun than the Earth which, moving faster in its smaller orbit, sometimes overtook them, making them look, from our point of view, as though they were travelling in reverse.

Perhaps unsurprisingly, this was not a popular suggestion, particularly with the Church. No one at that time made an enemy of the Church as the consequences of doing so could be grim. *Revolutions* was published with an introduction (added without Copernicus' approval) that these revolutionary ideas need not be considered as true. In 1616, *Revolutions* was placed on the Catholic Church's list of banned books, where it remained until 1835.

Nicolaus Copernicus

Word of the Copernican model spread slowly. Copernicus still believed the universe was formed of perfect spheres, they just no longer centred on the Earth. Then, at the beginning of the 17th century, the German astronomer Johannes Kepler's painstaking observations led him to a sensational conclusion. The paths of the planets were not perfect circles, they were flattened circles, or ellipses. After Galileo's discovery of the moons of Jupiter, Kepler found that these, too, moved in elliptical paths around the giant planet.

Kepler set out his three laws of planetary motion; these described how the planets moved, but not why they moved. He tried to work out what force might be responsible for the planets moving as they did. Kepler thought magnetism might be involved and that the Sun must have something to do with it, but couldn't arrive at a satisfactory explanation. This would not emerge until 50 years later, with Isaac Newton and his ideas about gravity.

Until Einstein, our understanding of the laws that govern the movement of objects through space was founded on the work of the British scientist Isaac Newton (1643–1727). The oft-repeated story of Newton in the apple orchard is familiar to every schoolchild and perhaps has lost its impact over the years, but it took a singular mind to ask the question: 'Why doesn't the Moon fall to the Earth like the apple does?' And it took real genius to conclude that the Moon *is*, in fact, falling.

The Universal Force

Newton knew that any theory he devised to describe the motion of both the apple and the Moon would have to explain Kepler's findings too. In 1687, he produced what is considered by many to be one of the greatest works of science ever written. *Philosophiae Naturalis Principia Mathematica* (*Mathematical Principles of Natural Philosophy*), often referred to simply as the *Principia*, set out Newton's vision of a universe in which all events take place against a backdrop of infinite space and smoothly flowing time.

Building on Galileo's experiments carried out on moving objects and Kepler's observations of the planets, Newton set out his three laws of motion and his theory of gravity.

NEWTON'S LAWS OF MOTION

1: An object will remain at rest or continue to move in the same direction and at the same speed unless acted on by a force.

2: A force acting on an object will cause it to move in the direction of that force. The magnitude of the change in the speed or direction of the object is dependent on the size of the force and the mass of the object.

3: For every action there is an equal and opposite reaction. If one object exerts a force on another, an equal and opposite force is exerted by the second object on the first.

Isaac Newton examining the nature of light with the aid of a prism.

Newton determined that between any two objects there is always a gravitational force which attracts them to each other. The strength of the force depends on the mass of each of the objects and on the distance between them. Gravity obeys an inverse square law, which means that the magnitude of the force decreases by the square of the distance. Therefore, if you double the distance between two objects, the force that draws them together reduces to just a quarter of what it was. At five times the distance, the force is reduced to a 25th of what it was.

With three simple laws of motion and one law of gravity, Newton, it seemed, could explain the movement of everything in the universe. His laws provided an explanation for Kepler's laws of planetary motion and for the fall of the apple. Newton derived his laws from three fundamental quantities which underpin all of science – time, mass and distance. By knowing the time an object takes to travel a set distance, it is possible to calculate its velocity (speed and direction). Mass tells us how much matter the object contains and therefore the amount of force required to move it. Multiplying the mass by the velocity gives the object's momentum, which indicates how difficult it will be to stop the object once it is moving. Later, Einstein would reveal that all three of these quantities are relative.

Absolute time and space

According to Newton, time and space were absolute; they were the stage upon which the drama of the universe unfolded, remaining unchanged by events. Newton thought of our everyday measures of the passing of time – the hour, the month, the year – as simply common time. Although useful, they were not to be confused with 'true', or 'absolute', time, as Newton called it. Absolute time, he believed, was completely separate from space and independent of events. Absolute time ticked along at the same steady pace throughout the universe.

One second for you should be exactly the same as one second for me, wherever we were in the universe and whatever we were doing.

Newton also believed in the idea of absolute space. He thought it should be possible to state the absolute position of an object in absolute space, almost like covering the universe

COSMIC CANNONBALLS

Two forces govern a cannonball's path – gravity and the force that propelled it from the cannon. The result of those two forces acting on the cannonball is that it follows a curved path back to Earth. Imagine that the cannon produced sufficient force that the cannonball's path matched the curvature of the Earth. It would now travel right around the Earth, always falling around the planet but never reaching the ground. (Let's assume for the sake of the argument that there's no air resistance to slow it down.) The cannonball is now a satellite in orbit. This is exactly the principle that puts real satellites into orbit, except with powerful rockets rather than a cannon providing the forward motion. The Moon is like a cosmic cannonball, perpetually falling around the Earth in its orbit.

in three-dimensional graph paper and plotting the positions of everything in it. But it is no more possible to say what absolute space is, as it is to define absolute time.

Newton's laws went unchallenged for more than 200 years. For everyday purposes, they are still an excellent way of calculating the movement of an object and how it is affected by gravity. But Newton had failed to explain what caused gravity. When Einstein came along he offered a suggestion that might have astonished even Newton.

CHAPTER 2

What is light?

The nature of light underpins much of Albert Einstein's work.

For centuries, people have tried to explain the various phenomena associated with light. The Ancient Greek philosopher Pythagoras believed that sight was like a very delicate sense of touch, the eyes producing invisible rays with which we sense objects. Another Ancient Greek thinker, Democritus, believed that objects continually emitted images of themselves, which we sensed.

The obvious drawback to both these ideas was that if this was the case then why isn't it possible for humans to see well at night? Plato put forward the idea that the inner light from the eyes had to mix with the light from the Sun before we could see anything. Aristotle suggested that we could only see things if they were illuminated, but this notion was rejected as too simple!

Waves or particles?

Whatever they believed the origins of light to be, the Ancient Greeks had opposing views concerning its nature. The first was that light is a disturbance in the ether, which was an invisible, undetectable substance that filled space. Aristotle thought light was a wave that travels through the ether like an ocean wave travelling

Pythagoras, 6th century BC mathematician and philosopher

through water. Another view held that light was a stream of tiny particles which were too small and fast-moving to be perceived individually. Plato and Aristotle both opposed the particle theory so, for the next 2,000 years, it was generally accepted that light travels in waves.

Let there be light

The Arab physicist Alhazen (965–1038) finally laid to rest the idea that beams of light emanated from the eyes. He established that we see things either because they reflect light from a source of illumination, or because they themselves are a source of illumination, whether it be a candle or the Sun.

The English scholar Robert Grosseteste (c.1168–1253) read Alhazen's work and carried out some experiments of his own. He believed the entire universe had been formed from light, which was the first of all things to be created, expanding out from a single point into a sphere that contained all other things within itself. This startlingly sophisticated notion had obvious parallels with our current thinking of how the universe formed.

Particles or waves?

In addition to his work with motion and gravity, Isaac Newton was also fascinated by light. He carried out many experiments and had his own ideas about its nature. He demonstrated that white light can be split into a rainbow of colours – a spectrum – by passing it through a prism. He noticed that light travels in straight lines and that shadows have sharp edges. It seemed obvious to Newton that light was a stream of particles and not a wave.

The British polymath and physician Thomas Young (1773–1829) was a man of such formidable intellect that his fellow students at Cambridge University dubbed him 'Phenomenon'. He had different ideas on light and decided to tackle the wave/particle problem by experiment. He theorized that if the wavelength of light was sufficiently short then it would appear to travel in straight lines, as if it were a stream of particles. In 1803, he carried out an experiment that was beautiful in its elegance and simplicity.

He began by making a small hole in a window blind, which gave him a point source of illumination. Next, he took a piece of board and made two pinholes in it, placed close together. He positioned his board so that the light coming through the hole in the window blind would pass through the pinholes and on to a screen. If Newton was right, and the light was a stream of particles, then there would be two points of light on the screen where the particles travelled through the pinholes. So, what did Young see? Rather than two discrete points of light, he saw a series of curved, coloured bands separated by dark lines, exactly as would be expected if light were a wave.

Waving goodbye to particles . . . for now

Two years earlier, in 1801, Young had described an effect he called 'interference'. If two waves meet, they don't bounce off each other like colliding snooker balls, instead they appear to pass straight through each other. Observe raindrops falling on to the surface of a pond and see how the ripples spread and meet and keep on going as they cross one another.

Where the waves cross they combine with one another. If the peak of one wave meets the peak of another, they are added together to make a higher peak; two troughs make a deeper trough and a trough and a peak cancel each other out. The result was an interference pattern that showed where the waves were adding and cancelling. And this was what Young had observed on his screen.

Unfortunately, it was not acceptable to contradict the great Isaac Newton, and Young's findings were not well received. But still the question remained: What actually was light? And if it was a wave, how did it travel through space? The beginnings of an answer came from the study of a seemingly unrelated force – electricity.

Secrets of electromagnetism

As the 19th century progressed, scientists gained an increasing amount of knowledge about electricity. Among their discoveries was the intimate relationship between electricity and magnetism.

In 1820, the Danish physicist Hans Christian Øersted discovered that a wire carrying an electric current would deflect the needle of a compass. The French scientist André-Marie Ampère (who gave his name to the amp) experimented further. He found that if he placed two wires, both carrying an electric

André-Marie Ampère

current, close to each other, one of two things happened. If the current was flowing in the same direction in each wire, the wires were pushed apart. If the current was flowing in opposite directions, the wires were drawn together. They were, in other words, behaving just like magnets.

Electromagnetic induction

In the course of hundreds of experiments carried out at the beginning of the 19th century, the English scientist Michael Faraday (1791–1867) discovered that just as an electric current produced magnetism, so a magnet moving through a coil of wire generated electricity. The magnet had to be moving – if it was stationary nothing happened. Electric currents make magnetic fields and moving magnets produce electric currents, a process that is now called electromagnetic induction. That magnetism and electricity were related on some fundamental level could no longer be in doubt. Our world would be a very different place without this discovery. It is the principle behind all the electricity generated in the world's power stations and what drives all the countless electric motors we use.

The Faraday effect

Faraday was convinced there was a link between electricity, magnetism and light. In 1845, he carried out an experiment in the basement of the Royal Institution in London and found he could affect the polarization of a beam of light using an electromagnet. This demonstrated that light did indeed have magnetic properties. Faraday wrote in his notebook: 'I have at last succeeded in . . . magnetizing a ray of light'. This demonstration of what is now called the 'Faraday effect' was an important stepping stone in Faraday's development of the field theory of electromagnetism which, in turn, would influence the work of the Scottish physicist and mathematician James Clerk Maxwell (1831–79) and, later, Albert Einstein.

Fields and forces

Faraday wanted to explain how a magnet could induce an electric current in a wire without coming into physical contact with it. To do so, he came up with the idea of an electromagnetic field. He saw this as lines of force, which he called 'flux lines', stretching invisibly through all of space. It's easy to make these lines visible – simply place a magnet under a sheet of paper and scatter some iron filings on the paper. The patterns formed by the filings reveal the magnetic lines of force. According to Faraday's field theory, the magnet was not the centre of the magnetic force, but concentrated the force through itself. The magnetic force wasn't in the magnet, but in a magnetic field in the space surrounding it.

Towards a deeper understanding

Some 20 years after Faraday proposed his field theory, the idea was taken up by James Clerk Maxwell, who set out to express Faraday's ideas mathematically. Einstein would later describe Maxwell's work on electromagnetism as 'the most profound and most fruitful that physics has experienced since Newton'.

In four equations, Maxwell detailed the electric and magnetic phenomena observed and recorded by Faraday and other researchers. Maxwell's equations described how electric charges and currents create electric and magnetic force fields. They demonstrated how an electric field can generate a magnetic field, and vice versa. Maxwell also proposed the existence of electromagnetic waves, which can propagate vast distances through empty space. His third and fourth equations described the way in which fields circulate around their respective sources. Imagine an electromagnetic wave as two field waves (one electric and the other magnetic) travelling in the same direction but perpendicular to each other. Both fields oscillate and keep in step with each other as the wave travels along. Maxwell saw that electricity and magnetism were always bound together – it was impossible to have one without the other.

He used his equations to calculate the velocity of an electromagnetic wave. The answer he obtained was 299,792,458 metres per second (m/s). This confirmed what experiments had shown to be the velocity of light. Maxwell thought this couldn't possibly be a coincidence and declared that light itself was an electromagnetic wave.

THE ELECTROMAGNETIC SPECTRUM

Maxwell predicted there should be a whole range, or spectrum, of electromagnetic waves, and so it proved. Infrared and ultraviolet light, invisible to human eyes, had already been discovered at either end of the visible spectrum and had the same wave-like properties as visible light. After Maxwell's death, the discovery of long wavelength radio waves and very short wavelength X-rays and gamma rays extended the spectrum further.

CHAPTER 3

How does light travel through space?

If light is a wave, how does it travel through the vacuum of space?

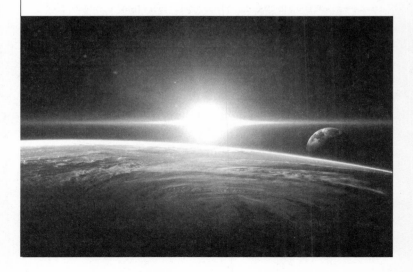

Waves need a medium to carry them. You can make waves in the water of a swimming pool by moving your arms up and down; if you clap your hands, you can send a sound wave through the air. But how does the light from the Sun travel from there to here? It was the view of James Clerk Maxwell and his contemporaries that light must also travel through a medium. They called it 'ether'.

A ghostly substance

Ether appeared to be undetectable and seemingly offered no resistance to planets or any other objects passing through it. Light apparently passed through it undiminished and failed to illuminate the ether in any way. Yet ether must fill all of space, since the light from the stars reaches us from every direction. So what was this ghostly substance which filled the universe?

The ingenious experimenters of the 19th century set out to find a way to pin down the elusive ether. Two particularly dedicated investigators were the American scientists Albert Michelson and Edward Morley, who embarked upon a series of precise experiments to demonstrate the effects of ether on the light that passed through it.

Against the wind

As the Earth moved in its orbit around the Sun, the flow of the ether across the Earth's surface could, it was believed, produce an 'ether wind'. A beam of light travelling through the ether should travel faster if it was moving with the wind and slower if it was going against it. The aim of Michelson and Morley's crucial experiment, carried out in 1887, was to measure the speed of light in different directions and by this method determine the speed of the ether relative to the Earth.

To carry out the measurements, Michelson designed a device called an interferometer. This sent the beam from a light source through a half-silvered mirror, which split the light into two beams travelling at right angles to each other. The beams were then reflected back to the middle by two more mirrors. The beams recombined, producing an interference pattern that could be observed through an eyepiece. Any variation in the time it took for the beams to travel between the mirrors would be seen as a shift in the interference pattern. The interferometer floated in a trough of mercury, which allowed it to be rotated slowly. If the ether theory was correct, the speed of the light beams would change as their direction altered in relation to the direction of the Earth's orbit.

Michelson and Morley discovered it made not one bit of difference how they rotated the apparatus. Nor did it matter at what time of day they took their measurements. The speed of the light beams was always the same. It seemed that the ether didn't exist after all.

Michelson's interferometer

Searching for an explanation

The world of physics was shaken by the failure to find evidence for ether. No one doubted the reliability of the experiment, but there was reluctance to accept its result. Physicists looked for a way to explain the findings and still corroborate the existence of ether. Michelson was perplexed by the results. He repeated the experiment, even trying it on a mountaintop, but the speed of light remained the same – there was not the slightest indication that the ether existed.

Michelson even wondered if this was because the ether stuck to the Earth and was dragged around with it.

The Lorentz–FitzGerald contraction

Working independently, Dutch physicist Hendrick Lorentz and Irish physicist George FitzGerald came up with the same solution. In 1889, FitzGerald published a short paper in which he proposed that the results of the Michelson–Morley experiment could be explained only if objects reduced in length as they travelled through the ether.

The reduction in length suggested was infinitesimal, amounting to just a couple of centimetres for an object the size of the Earth, but it would be enough to explain the results. A few years later, a clerk working in a Swiss patent office showed that the whole idea of ether was simply unnecessary. Albert Einstein suggested that all you needed to do was give up the notion of absolute time.

HOW DO ELECTROMAGNETIC WAVES TRAVEL THROUGH SPACE?

An electromagnetic wave is produced by changes in the electric and magnetic fields. As demonstrated by Faraday and others, a changing electric field produces a changing magnetic field, and vice versa. An electromagnetic wave is self-propagating and doesn't need a medium to travel through. Unlike an ocean wave displacing water molecules as it passes through them, an electromagnetic wave displaces nothing in space. You might think of an electromagnetic wave as an energy-carrying disturbance travelling invisibly through space, until it interacts with matter.

CHAPTER 4

What is the quantum?

Max Planck's radical idea – the quantum – transformed physics and laid the groundwork for Einstein's own theory.

If you heat a metal rod sufficiently it begins to glow – it gets red hot. If you increase the heat it glows yellow and eventually white hot, giving off light at all wavelengths of the spectrum. Why should this be? Why does increasing the temperature cause electromagnetic waves to be produced?

Blackbody radiation

Every object gives off electromagnetic radiation, called 'blackbody radiation', all the time; the amount of radiation depends on the object's temperature. A blackbody is simply something which absorbs the electromagnetic radiation that hits it and then radiates it out again, mostly in the form of infrared radiation which we detect as heat. The hotter an object is, the more energetic the blackbody radiation it emits. If it is hot enough, it will pump out blackbody radiation in the form of light and higher frequency electromagnetic waves.

This was a problem for classical physicists and their understanding of how electromagnetic waves worked. It made sense that the light should get brighter, but why would it change colour?

In theory, an ideal blackbody would absorb and emit radiation at all frequencies, but nothing in the real world is ideal. All the objects in the universe are exchanging electromagnetic radiation with one another all the time. This is why nothing will ever cool to absolute zero, the theoretical lowest possible temperature at which a substance transmits no energy at all.

It's a catastrophe!

When the physicists of the late 19th century tried to explain their observations of blackbody radiation they ran into a problem. According to the laws of physics as they were understood at the time, a hot body should give off radiation at all frequencies, including short wavelength X-rays and gamma rays, and longer wavelengths such as radio waves. Since there is no upper limit on the higher frequencies, and therefore since there are many more higher frequencies than lower ones, this would mean an infinite number of waves being generated, all of them carrying energy. This came to be known as the ultraviolet catastrophe.

There had to be a flaw somewhere in the ideas about thermodynamics and electromagnetism, but what was it? No one could provide an answer. Then the German physicist Max Planck proposed a radical and stunning solution which transformed physics.

Into the quantum universe

On 19 October 1900, Planck gave a speech to the German Physical Society that heralded the beginning of a new age for physics. He suggested that instead of being a continuous and infinitely variable quantity like a wave, energy comes in packets. He called these packets quanta (the plural of quantum) from the Latin meaning 'how much' or 'how many'. A quantum was defined as the minimum amount of any physical property involved in an interaction.

According to Planck, energy could only be emitted or absorbed in whole quanta, and each quantum had its own corresponding wavelength and frequency. This explained why a blackbody would not give off radiation equally across the whole electromagnetic spectrum.

The ultraviolet catastrophe was prevented from happening because even though there are many more higher frequencies than lower ones, increasingly large amounts of energy were required to achieve them. A quantum of violet light, for example, has twice the frequency and therefore twice the energy of a quantum of red light.

As they heat up, objects progress through the spectrum, from infrared to red, orange, blue and white hot, because the increase in temperature means that their constituent particles are becoming more energetic. This increase in energy is what allows higher frequency quanta to be formed. Planck suggested that the energy of a quantum was related to its frequency. He used the formula $E = \hbar v$, where E equals energy, $\hbar$ equals a value and v equals frequency – this is known as Planck's constant. The energy of a quantum of radiation can be calculated by multiplying its frequency by Planck's constant. Using his formula, Planck could accurately predict the total amount of energy in an oven at any given temperature.

There could be no doubt that Planck's solution worked; what the theory predicted agreed with what was found by experiment. But Planck thought his own explanation was unlikely because it contradicted everything he had been taught. Even so, he accepted it as a convenient solution for what happened when matter absorbed or emitted energy, even though he was unable to give a good reason why it should be true. A few years later, Albert Einstein reinterpreted Planck's hypothesis and expanded on the theory.

WHAT IS A CONSTANT?

In physics, certain fundamental quantities appear that always remain the same whatever the conditions. These are called 'physical constants'. They include Planck's constant, h, linking energy and frequency in a quantum; c, the velocity of light, and G, the gravitational constant. Physical constants, it is believed, hold true under all circumstances and do not change.

While Planck was presenting his theory, the 21-year-old Albert Einstein was working as a maths tutor and writing a paper on the capillary effect. Later he described his reaction to Planck's theory: 'It was as if the ground had been pulled out from under us, with no firm foundation to be seen anywhere.' Planck and Einstein spent a great deal of time together, exchanging ideas over many years. When Planck died in 1947, Max Born remarked: 'It is difficult to imagine two men of more different attitudes to life. . . . Yet what did all these differences matter in view of what they had in common – the fascinating interest in the secrets of nature.'

FIRING OUT PHOTONS

When atoms absorb sufficient energy, the electrons surrounding the atomic nucleus can jump into higher orbits. As the electron falls back to its original orbit it releases a photon – a quantum of electromagnetic energy. The amount of energy in the photon depends on how far back the electron falls. The further from the nucleus an electron is boosted, the higher the energy of the photon that is released.

CHAPTER 5

What is the photoelectric effect?

Einstein took Planck's theory of the quantum and used it to explain a puzzling phenomenon.

Another curious aspect of radiation – the 'photoelectric effect' – needed an explanation. It was already known that if a beam of light was directed at certain types of metal, electrons would be emitted. This is the science which underpins solar-generated electricity.

At first it seemed the effect could be explained in terms of electromagnetism. It was assumed that the electric field part of the electromagnetic wave gave the electrons the energy they needed to break free from the metal. But it soon became apparent that this couldn't be the whole story.

The energy of the electrons released depended on the frequency of the light – not on its intensity. It would seem obvious that a brighter light should have more energy and therefore produce higher energy electrons but, no matter how bright the light used, the electrons which emerged still had the same energy. It was only by shifting the light up in frequency, from red to violet and then ultraviolet, that higher energy electrons were emitted. A brighter light produced more electrons, but their energy stayed the same. If the light was of a low enough frequency no electrons would be emitted at all, even if the light was blindingly bright. The wave theory of light failed to explain these findings.

Light quanta

Einstein was intrigued by the phenomenon. In 1904, he wrote to a friend that he had 'found in a most simple way the relation between the size of elementary quanta of matter and the wavelengths of radiation'. In a paper he published in the journal *Annalen der Physik* (*Annals of Physics*) in March 1905, Einstein took the findings of the experiments on the photoelectric effect and married it to Planck's theory, producing a result which would eventually win him the 1921 Nobel Prize.

Describing light as packets of energy, Einstein wrote: 'When a light is propagated from a point, the energy is not continuously distributed over an increasing space but consists of a finite number of energy quanta which are localized at points in space and which can be produced and absorbed only as complete units.' Einstein's biographer Walter Isaacson described this as 'perhaps the most revolutionary sentence that Einstein ever wrote'.

Einstein's approach was simple yet compelling. He compared the formulae which described how the particles in a gas behaved as it changed volume with those that described similar changes as radiation spread through space. He found that both obeyed the same rules. The maths underpinning the behaviour of the gas was the same as that for the radiation. It allowed Einstein to calculate what the energy of a light quantum of a particular frequency would be – his results agreed with Planck's findings.

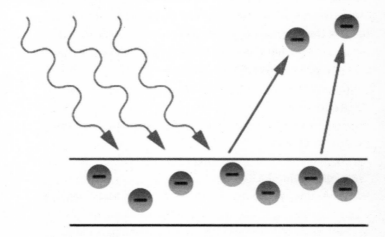

The photoelectric effect

The photoelectric effect explained

Einstein then demonstrated how the existence of light quanta explained the photoelectric effect. As Planck had shown, the energy of a quantum was determined by multiplying its frequency by Planck's constant. If a single quantum were to transfer all its energy to a single electron, then the energy of the emitted electron would be equal to that of the quantum. Increasing the intensity of the light would produce more electrons, but wouldn't increase their energy. This was in line with laboratory observations of the photoelectric effect.

Quantum reality

As far as Planck was concerned, the quantum was little more than a mathematical contrivance which he used to make the equations work out. But to Einstein it was a physical reality, a feature of the universe as it really was. By 1916, experiments had confirmed that Einstein was right and scientists were having to rethink their ideas on the nature of light.

For the next 20 years, Einstein struggled without success to solve the dual nature paradox of light. Towards the end of his life, in 1951, he wrote in a letter to his friend Michele Besso that 50 years of investigation had brought him no closer to finding an answer to the question, 'What are light quanta?' 'Nowadays every Tom, Dick and Harry thinks he knows it,' Einstein wrote, 'but he is mistaken. . . . No one really knows exactly what light is!'

Wave–particle duality

What became apparent was that, bizarre and mind-boggling as it might be, the wave theory and the particle theory of light are both correct. Whether light is a wave or a particle seems to depend upon how you look at it. All we can do is describe how light behaves under different conditions – sometimes it

QUANTUM STRANGENESS

Thomas Young (see page 29) had demonstrated that light was a wave by showing how light passing through two close-together slits formed interference patterns. But supposing that light is a stream of particles, what would happen if we sent just one quantum at a time towards the slits? You'd imagine that the interference patterns would be replaced by two bright patches in line with the slits. But, spookily, even if we only fire off one photon a second, the interference patterns still form. Even if subsequent photons are fired off after the earlier ones have hit the screen, they still somehow 'know' where to go to build up the interference pattern! How does that work? As the great physicist Richard Feynman said: 'A lot of people understood the theory of relativity in one way or another. . . . On the other hand, I can safely say that nobody understands quantum mechanics.'

Richard Feynman

behaves like a wave and sometimes it behaves like a stream of particles. Sometimes it seems to be both at the same time. We have no single model to describe light in all its aspects. It's easy enough to say that light has 'wave–particle duality' and leave it at that, but what that actually means is something no one can answer satisfactorily.

CHAPTER 6

How did Einstein prove that atoms exist?

At the beginning of the 20th century scientists were still arguing about whether atoms and molecules existed. Einstein took the work of a Scottish botanist and showed that they did.

In May 1905, the *Annalen der Physik* received another paper from Einstein. The subject this time – the kinetic theory of gases – was grounded in classical physics, but Einstein would use it to establish for the first time that atoms and molecules were a physical reality.

Which way are we going?

Earlier we looked at Newton's laws of motion, which allows us to calculate the path of a comet-catching space probe or the flight of a cricket ball into the wicketkeeper's gloves. Using Newton's laws, if you know how an object is moving at the moment, you can work out how it was moving in the past, and how it will move in the future.

The funny thing about the laws of motion is that they aren't time dependent. If you look at a moving image sequence of a comet flying through space, there is no way Newton's laws can tell you the direction in which the film is running. Newton's laws are time reversible – they work just as well in either direction, whether you are calculating motion forwards, into the future, or backwards, into the past.

For most situations, common sense and experience tell us whether an event is taking place forwards or backwards in time. If I showed you a film of a broken egg putting itself back together and rising up off the floor and into my hand, you'd know at once that I was running it in reverse. Newton's laws don't say this is impossible, but it's fairly certain you'll never see it happen because it is highly unlikely.

The reversibility paradox

According to the kinetic theory of gases, heat is a measure of the motion of atoms. The more agitated the atoms, the higher the heat. The Austrian physicist and philosopher Ludwig Boltzmann used the kinetic theory to resolve the so-called 'reversibility paradox' in physics. This arose from the second law of thermodynamics, which states that physical systems tend to become more disordered and dictates that most natural processes are irreversible. The universe, it seems, proceeds inexorably from a state of low entropy (order) to a state of high entropy (disorder), and this appears to contradict the time-reversible nature of Newtonian mechanics. The idea of an 'arrow of time', pointing from the past to the future, was first introduced by the astronomer Sir Arthur Eddington.

Time reversal symmetry versus the arrow of time

Boltzmann solved the paradox by determining that the second law concerned probabilities. The countless atoms and molecules that make up a broken egg or any object are in constant random motion. There is a vanishingly small but not impossible chance that the molecules will all move in exactly the right direction to reassemble the egg. But this event is so improbable that the breaking of the egg is irreversible.

Einstein read Boltzmann's theory of gases – which portrays a gas as a collection of countless molecules bouncing around randomly – and declared it 'absolutely magnificent'. Between 1902 and 1904, Einstein was also working on the second law of thermodynamics and developing a 'general molecular theory of heat' using statistics and mechanics that would 'close the gap', as he put it, and extend Boltzmann's work on gases to other materials. He presented his statistical molecular theory for his doctoral dissertation at the University of Zurich, describing a new theoretical method for determining the size of molecules and for calculating Avogadro's number – the number of atoms or molecules in a set amount of a substance.

$$N_A = 6.02 \times 10^{23}$$

Avogadro's number, or the Avogadro constant, is named after Italian scientist Amedeo Avogadro who, in 1811, first proposed that the volume of a gas is proportional to the number of its atoms or molecules.

In a separate paper, published in May 1905, Einstein applied the molecular theory of heat to liquids and ended up solving the puzzle of 'Brownian motion'.

Brownian motion

In 1827, the Scottish botanist Robert Brown noticed that flower pollen suspended in water moved about randomly, seemingly being nudged by unseen forces. Other researchers had noticed this odd phenomenon, but Brown was the first to study it. To begin with, Brown thought it had something to do with the pollen being alive, but experiments showed that not only pollen grains moved around in this way. Any particles of a similar size, from chips of granite to particles of smoke, if suspended in liquid, showed the same odd motion.

Einstein didn't specifically set out to explain Brownian motion and didn't even mention it in the title of his paper, calling it 'On the Motion Required by the Molecular Kinetic Theory of Heat of Particles Suspended in Fluids at Rest'. He wrote: 'It is possible that the motions to be discussed here are identical with so-called Brownian molecular motion; however, the data available to me on the latter are so imprecise that I could not form a judgement on the question.'

Einstein wanted to find evidence for the existence of atoms and molecules and to show how the actions of molecules could be visibly demonstrated. Explaining Brownian motion was incidental to that. Einstein reasoned that if the molecules in a liquid were moving randomly, like the molecules in a gas, then every so often they would jostle against the tiny pollen grains in sufficient numbers to make them move. Einstein explained the motion in detail, using theoretical knowledge and the

Brownian motion

| Solid | Liquid | Gas |

The three fundamental states of matter – the molecules of a solid, liquid and a gas in a glass beaker.

data from experiments, coupled with the powerful statistical tools he had mastered in his earlier dissertation, to accurately predict how far the particles would travel in the course of their irregular, random motions.

When Einstein's paper on Brownian motion first appeared in 1905, scientists were still debating the very existence of atoms and molecules. Some scientists such as the physicist Ernst Mach (who gave his name to the speed of sound) and the physical chemist Wilhelm Ostwald were among those who chose to argue against the atom. They took the view that thermodynamics concerned the way in which energy changes from one form to another and believed there was no need to explain it in terms of randomly moving invisible atoms.

Mach had an influence on Einstein's thinking in another way. He declared it was impossible to define Newton's concepts of absolute time and space, calling it a 'conceptual monstrosity'. But Einstein would overturn these ideas altogether. Within months of the publication of Einstein's paper, his predictions had been confirmed by experiment.

The French physicist Jean-Baptiste Perrin used the newly invented ultramicroscope to verify Einstein's ideas and was awarded the 1926 Nobel Prize for physics for his work in this field. He found compelling evidence to support Einstein's theory. The German physicist and mathematician Max Born wrote: 'I think that these investigations of Einstein's have done more than any other work to convince physicists of the reality of atoms and molecules.'

It is a mark of Einstein's genius that while he was proving the existence of atoms he was also working out the consequences of travelling at the speed of light. Just a few days after the publication of his paper on molecular motion, he was telling a friend that he was about to modify 'the theory of space and time'.

CHAPTER 7

What is the theory of special relativity?

Einstein showed that because the speed of light is constant, everything else had to change.

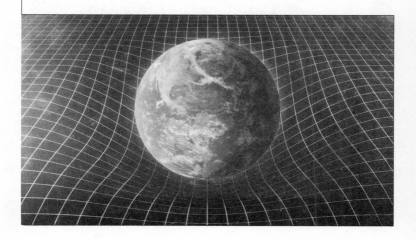

The idea of relativity in physics is a fairly straightforward one. It maintains that the laws of physics apply and remain the same for all freely moving observers, whatever their speed of motion. But what do we mean by 'being in motion'?

Einstein's theory of special relativity, developed in 1905, was special in the sense that it concerned the special state of objects in uniform motion, moving at a constant velocity and direction relative to one another. Physicists call this an inertial frame of reference. As Newton pointed out in his first law of motion, an inertial state is the default for any object not being acted on by a force. Inertial motion is simply motion at a uniform speed in a straight line. It would be another ten years before Einstein formulated his general theory, which also encompassed objects in accelerated motion.

To measure anything, be it time or distance or mass, it's necessary to have something to measure it against. An object is only faster, bigger or heavier relative to something else.

One reason for developing a system of weights and measures was to have a way to compare similar objects and quantities in order to establish that one thing was bigger or heavier than another. None of our units of measurement are absolute; they all have to be defined by reference to something else.

Galileo

In 1632, Galileo Galilei had explored the idea that all motion is relative. In his *Dialogo sopra i due massimi sistemi del mondo* (*Dialogue Concerning the Two Chief World Systems*), Galileo defended the idea that the Earth does not sit motionless at the centre of the universe. When Copernicus had suggested that the Earth moves around the Sun, his idea had been ridiculed by the critics who said that we would surely feel it rushing through space.

Galileo tackled the problem by imagining the scenario of a person inside a windowless cabin on a ship sailing on a perfectly smooth lake at a constant speed. Galileo asked: is there a way in which the passenger can determine that the ship is moving without going on deck? If the ship continues to move at constant speed and direction, the passenger will not feel its motion. Similarly, a modern-day passenger travelling on a smoothly running train or in an aircraft cruising through the sky will not be aware of the speed of travel without looking out of the window.

Galileo wondered if an experiment could be carried out on the ship that would give a different result to the same experiment carried out on the shore, and therefore indicate whether the ship was in motion. He concluded that it was impossible. Provided that the ship was moving at a constant speed in a constant direction, any mechanical experiment performed inside it would give exactly the same results as a similar experiment carried out on shore. From these observations, Galileo put forward his relativity hypothesis:

Any two observers moving at constant speed and direction with respect to one another will obtain the same results for all mechanical experiments.

Galileo Galilei

Frames of reference

An important consequence of this is that velocity can only be measured with reference to something else, and the calculation changes if we measure the velocity from a different reference point. To say something is moving only has meaning if you can say what it is moving in relation to – if two people sit opposite each other on a train and one tosses the other an orange, it travels through the air at a few kilometres per hour, but someone standing at the side of the tracks would see orange, train and passengers rush by at a hundred kilometres per hour! The speed at which you perceive an object to be moving depends on how fast you yourself are moving relative to that object.

The idea that motion has no meaning without a frame of reference is fundamental to Einstein's theories of relativity. Before Einstein, it was believed that there was such a thing as absolute motion; this means that an object could be said to be moving without reference to anything else. This required there to be a state of absolute rest in space (either something is moving or it isn't). This idea was formulated by Newton, who wrote: 'Absolute motion is the translation of a body from one absolute place into another; and relative motion, the translation from one relative place into another.' Einstein's theory of special relativity did away with this idea of absolute rest and absolute motion. On one (perhaps apocryphal) occasion, Einstein asked a bemused ticket inspector, 'Does Oxford stop at this train?'

Introducing special relativity

Einstein's third paper of 1905 was called 'On the Electrodynamics of Moving Bodies'. It started with a very simple example, well known to Michael Faraday and other Victorian researchers into electromagnetism: an electric current is generated if a magnet is moved inside a coil of wire, and the same current is produced if the magnet remains fixed and the coil is moved. Einstein was familiar with electricity – he often helped his engineer uncle Jakob, who was also responsible for introducing the young Einstein to the delights of algebra, to tinker with the coils and magnets in a generator. Einstein's job in the patent office also meant he was regularly examining a variety of electromechanical devices.

Since Faraday's time, it had been assumed that there were two different explanations at work – one for the moving magnet producing a current and another for the moving coil producing the current. But Einstein disputed this, saying it didn't matter which was moving, it was their movement relative to each other that generated the current. As he said, 'The idea that these two cases should essentially be different was unbearable to me.'

The distinction between moving magnet and moving coil depended on the view still held by most scientists that there was a state of absolute rest with respect to the ether, the mythical and mysterious substance for which Michelson and Morley had failed to find any evidence.

The example of the magnet and coil, along with other observations made about the nature of light, led Einstein to conclude that the idea of absolute rest was flawed and unnecessary. In a casual sentence, he dismissed the idea of the ether entirely: 'The introduction of the "light ether" will prove to be superfluous . . . the view developed here will not require a "space at absolute rest".' He set out his 'Principle of Relativity':

'*The same laws of electrodynamics and optics will be valid for all frames of reference for which the laws of mechanics hold good.*'

Another way of saying this is that the laws of physics are the same in all inertial frames of reference. No matter whether you're travelling fast or slow, this way, that way, forwards or backwards, the laws remain the same, meaning that any experiment will produce results that are in accordance with the laws. Einstein and Galileo are both in agreement that no experiment can determine the motion of the observer in an inertial frame.

Special relativity only applies to objects moving in an inertial frame. Once the object changes direction, or speeds up or slows down, it can be said to be in motion. We feel it when a car accelerates or a plane begins its descent. There is no need to say an object is accelerating with respect to anything else.

The constancy of light

Once Einstein had adopted his Principle of Relativity, he realized it was impossible for both Newton and Maxwell to be right. He began his challenge of 200 years of Newtonian physics, writing in 1940:

'*The precise formulation of the time–space laws was the work of Maxwell. Imagine his feelings when the differential equations he had formulated proved to him that electromagnetic fields spread in the form of polarized waves, and at the speed of light! To few men in the world has such an experience been vouchsafed . . . it took physicists some decades to grasp the full significance of Maxwell's discovery, so bold was the leap that his genius forced upon the conceptions of his fellow workers.'*

Einstein asked: Does light behave the same way as everything else? Is the speed of light also dependent on the motion of the observer? This brought him to the second postulate, called the light postulate, upon which he founded his theory – that the speed of light is a constant. Some things may be relative, but the speed of light is absolute.

Light, said Einstein, always travelled at a constant velocity that was independent of the velocity of the object emitting the light. This made little sense in Newtonian terms, in which speeds were added together. For example, a fast bowler can bowl a cricket ball faster by adding the speed of his run to the speed at which the ball is released from his hand. But a beam of light projected from a fast-moving aircraft would still travel at the same speed as one projected from a (relatively) stationary mountaintop below it.

This was what Michelson and Morley had found when they discovered that the speed of light was always the same no matter how they measured it. The idea that nothing in the universe can travel faster than light is central to Einstein's special theory of relativity. But why does light always travel through a vacuum at nearly 300,000 kilometres per second (km/s)? Why not faster or slower?

Simply put, it's because that's the answer we get when we solve Maxwell's equations, which show that the speed of electromagnetic waves is a constant defined by the properties of the vacuum of space through which the waves move. It

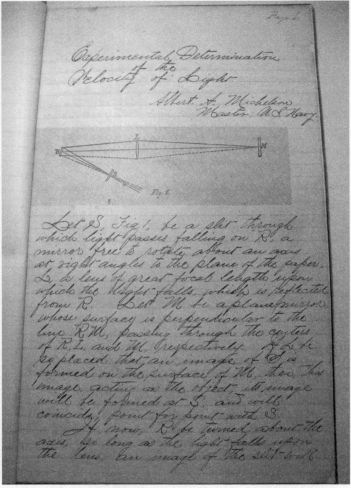

The first page of Michelson's handwritten draft paper on the velocity of light

is not measured relative to anything else as any other speed would be. The nature of the universe and the behaviour of electric and magnetic fields dictate what the speed of light must be.

Because Maxwell's equations, which determine the speed of light, hold true in any inertial frame, two observers moving relative to each other, each measuring the speed of a beam of light relative to themselves, will both get the same answer – even if one is moving in the same direction as the beam of light and one is moving away from it. Everything else in the special theory of relativity derives from that one simple fact. The constancy of the speed of light produces many seeming paradoxes that turn our notion of space and time on its head.

LIGHT SPEED DEFINED

In 1983, the General Conference on Weights and Measures officially defined the speed of light to be:

c = 299,792,458 m/s

Scientists use 'c' as the symbol for the speed of light from the Latin word 'celeritas', which means 'swiftness'. At the same time, the metre came to be defined as the distance light travels in one 299,792,458th of a second.

SLOW LIGHT

The 'speed of light' is generally taken to refer to light travelling in a vacuum. Light doesn't always travel this fast though – it slows down as it passes through transparent media such as air, water or glass. The speed of light through water is about 75 per cent of its speed in a vacuum, but that's still about 225,000 km/s so it's hard to detect the difference. The ratio by which it is slowed down is called the refractive index, discovered by Jean Foucault in 1850.

CHAPTER 8

What are Einstein's ideas about time?

The consequences of special relativity turned our notion of time on its head.

In a talk he gave in 1922, Einstein said:

> 'My solution was really for the very concept of time, that is,
> that time is not absolutely defined but there is an inseparable
> connection between time and the signal velocity [of light]. With
> this conception, the foregoing extraordinary difficulty could be
> thoroughly solved. Five weeks after my recognition of this, the
> present theory of special relativity was completed.'

Imagine you're in a space station and you fire a laser-
signalling device at a spacecraft that's moving away from you
at half the speed of light, around 150,000 km/s. Common sense
tells you that the laser light should reach the spacecraft at
half-light speed, because it has to catch up with the spacecraft.
But common sense is wrong – the beam will still arrive at the
spacecraft at approximately 300,000 km/s.

Velocity equals distance travelled divided by the time taken
to get there, or as an equation:

$$v = d/t$$

In other words, it is a measure of space divided by time. So it
seems that in order for the speed of light always to remain the
same for all observers (that is, that we always get the same
answer for v), we have to make some adjustments to d and t.
If light speed is unalterable, time and space must change.

Absolute time

Isaac Newton wrote that: 'Time exists in and of itself and
flows equably without reference to anything external.' In
Newton's world, time always ticked by at the same pace,
wherever you measured it. If our timepieces are both accurate,
then five seconds for me will be five seconds for you – except

they won't. Return to the space station and the laser signalling device for a moment. For you, on the space station, and the pilot in the spacecraft to agree on the speed the beam reaches the spacecraft, you both have to agree on the time it takes to get there. Since the speed of light is always the same for the spacecraft pilot, the spaceship's clocks have to be running slower for you to arrive at the same answer.

The twins paradox

Imagine that the pilot of the swiftly departing spacecraft, now accelerating to 0.99 per cent of light speed, is your twin who is setting out on a voyage of exploration through space. It will take your twin one year of ship's time to complete his or her journey. When your twin returns, he or she will be, as you'd expect, a year older. But how much older will you be?

As you track the ship's progress, you will notice something strange occur. The clocks onboard the ship are ticking more slowly than the clocks on the space station (or perhaps it's that yours are running faster, relatively speaking), so by the time your twin's ship returns to the space station, seven years will have passed for you. (The actual time will depend on just how close to light speed the spaceship travels – at half light speed, an hour on the ship would be equivalent to 69 minutes on the space station; very close to light speed, the difference might be measured in thousands or even millions of years.) You now have a twin who is six years younger than you – but that isn't the paradox.

Einstein said that all motion is relative. It didn't matter whether the coil or the magnet moved in order to produce a current; either way, they were moving relative to each other. By the same token, your twin aboard the spaceship might look back at the space station as it falls far behind and say it's you, instead, who is retreating at close to light speed. If this is the case, your twin will see your clocks moving more slowly! So, which of you is ageing more slowly – both of you, or neither?

You might think it all balances out and you and your twin will still be the same age when you're reunited. But you would be wrong. The spaceship pilot twin really will have aged less. Writing in the 1960s, physicist Herbert Dingle argued that the twin paradox revealed an inconsistency in special relativity. Today, most scientists agree that relativity theory can solve the puzzle, but they disagree about whether the solution can be found in special relativity. Einstein himself said the solution to the paradox requires general relativity.

Faster, and slower

According to special relativity theory, the faster you travel through space, the slower you travel through time. As you approach the speed of light, the intervals between events lengthen, so time seems to slow down. This phenomenon is called time dilation. If an object could achieve the speed of light, then time would appear to stop entirely. In experiments such as those carried out in the Large Hadron Collider at CERN, where atomic particles are smashed into one another at significant fractions of the speed of light, the effects of time dilation must be taken into consideration if the results are to make any sense.

Simultaneity

Einstein thought one effect of special relativity was particularly important – the relativity of simultaneity. Two events which appear to one observer to occur simultaneously might not appear to do so to a second observer, who is moving in relation to the first observer. According to Einstein, we cannot say one observer is right and the other wrong. In fact, they are both right!

Einstein explained the conundrum in terms of a thought experiment. Imagine you are watching a thunderstorm. Suddenly, two buildings you know to be equidistant from you are struck by lightning. You would say the two had been struck simultaneously. Now suppose a bus passes by; if the lightning strikes take place while a passenger on the bus is

THE NATURE OF TIME

Time can be long or short; it can hang heavy or fly by; we take time and we try to make time; we save time and we wonder where the time has gone. Nobody really knows what time is. In 1905, the French physicist Henri Poincaré argued that it is something we've invented for our own convenience rather than a feature of reality. He declared there were no tests that would tell us anything about the nature of time, and we should simply adopt whatever concept of time makes for the simplest laws of physics.

We might think of time as that which separates one event from another; it can tell us how long an event lasted and which event came first. For example, in a 100-metre race it takes a top sprinter ten seconds to run from the starting block to the finishing line. We can measure the time taken in fractions of a second so we know who won and whether or not it was a record time. But what is a second? We can define it in terms of the vibrations of an atom, but each vibration is just another event in time.

If nothing happened anywhere, if there were no events occurring at all, would time still 'flow equably', regardless? Perhaps time is just things happening. And, as Einstein showed, things can happen at different rates.

Henri Poincaré

level with you, he will doubtless agree that the two strikes occurred simultaneously. Now imagine that the bus is moving towards one building and away from the other. In that case, the light from the strike on the second building will take longer to reach the passenger than the light from the strike he is moving towards. He will not see the two strikes as simultaneous.

As we have seen, according to the principle of relativity you cannot insist that you are at rest and the passenger on the bus is in motion. You are simply in motion relative to each other. There is, therefore, no 'right' answer to whether or not the strikes occurred simultaneously.

The relativity of simultaneity demonstrates the impossibility of absolute time. Two observers in relative motion will have clocks that tick at different rates; the effect becomes more marked approaching the speed of light, but it is there, infinitesimally, at low relative speeds as well. Time passes differently for all moving reference frames.

As physicist Werner Heisenberg put it: 'This was a change in the very foundation of physics, an unexpected and very radical change that required all the courage of a young and revolutionary genius.'

CHAPTER 9

What is the Lorentz–FitzGerald contraction?

Not only did Einstein bend
time to fit his theory, he
shrank space too.

Another strange consequence of the speed of light remaining constant to all observers is the phenomenon that a moving object appears to shrink along the direction of motion. At the speed of light, the length of the object would be zero. This is the Lorentz–FitzGerald contraction, named after the two physicists who proposed it in 1889, following the failure of the Michelson–Morley experiment. It was Einstein who showed that the phenomenon was real – not an actual physical compression, but a consequence of the properties of space and time.

Just follow the bouncing beam

As time slows down the faster we move, it follows that we have to physically contract, too. Imagine a spacecraft with a mirror mounted on each end. A pulse of light bounces between these two mirrors. What happens to the bouncing beam as the spacecraft approaches the speed of light?

For a 150-metre-long ship at rest, the return journey for the light beam takes roughly a millionth of a second. But at 99.5 per cent of the speed of light, time is slowed down by around a factor of 10, which means the return journey time, as measured by an observer, is now a 100,000th of a second. With the pulse travelling from the back to the front of the spacecraft, the journey time will take longer because the front mirror is retreating from the pulse at close to the speed of light. With the pulse travelling from the front to the back of the spacecraft, the journey time will be much shorter because the rear mirror is rushing towards the light beam. But no matter whether it's a retreating or an advancing mirror, the light beam will always reach it at the same speed, roughly 300,000 km/s, because the speed of light doesn't change.

Einstein asked: If I could fly at the speed of light and held a mirror in front of myself, would I see my reflection? How would the light reach the mirror if it was retreating at light speed? The answer is that he would see his reflection because no matter how close he is to light speed, the light bouncing between him and the mirror and back again will always be travelling at the same 300,000 km/s.

To ensure that light speed is always measured to be the same, not only does time have to slow down, but the distance travelled by the light beam also has to decrease. At roughly 99.5 per cent of light speed, the distance is reduced by a factor of 10 – the same proportion as that of the time dilation effect.

Einstein with Dutch physicist Hendrik Lorentz

What is the Lorentz–FitzGerald contraction? | **77**

The spaceship and its crew don't shrink in size. The object in motion is only shortened in the direction of its motion; the dimensions that are perpendicular to its motion remain the same. The result is that, to an observer at rest relative to the moving object, it becomes distorted from its resting shape.

The change in length won't be apparent to the crew of the ship. The distortion will only be apparent to an observer who is relatively at rest compared to the ship. From the perspective of the ship's crew, the observer will appear to have contracted as, relatively speaking, he/she is streaking past them.

$v = 0$	$v = 0.87c$	$v = 0.995c$	$v = 0.999c$	$v \to c$
$L^* = L$	$L^* = 0.5L$	$L^* = 0.25L$	$L^* = 0.045L$	$L^* \to 0$

The effects of the Lorentz–FitzGerald contraction on a tennis ball

To boldly go

As the author Douglas Adams so memorably put it, 'Space is big. Really big. You just won't believe how vastly hugely mindbogglingly big it is.' Even light, the fastest thing in the universe, takes over four years to traverse the distance between the Sun and the next closest star. Yet should it ever become technically possible to attain near-light speed, travel to the stars for these swiftly moving pioneers would be much faster. How is that possible?

It's another consequence of the Lorentz–FitzGerald contraction. Imagine that the spaceship moving at light speed travels along a rail which stretches from star to star. The faster the ship travels, the shorter the rail will seem and the shorter the distance the spaceship will need to travel to reach its stellar destination. At 99.5 per cent of light speed, the journey to the nearest star will take about five months rather than four years. The closer the ship gets to the speed of light, the shorter the journey time will be.

There is a downside to this, of course. The ship's clocks are ticking ten times more slowly than the relatively motionless clocks of people back on Earth. So, although only five months will pass according to ship's time, by Earth time it will still take four years to complete the journey. The faster the ship travels, the more extreme the discrepancy between ship time and Earth time.

There in no time

Nothing can travel at light speed – except light, of course. So what would a journey across space be like for a photon, a single light quantum? Distances would shrink to zero and the photon's clock would stop ticking altogether.

For the photon, there is no distance and no time. A journey from one side of the universe to the other is accomplished in no time at all because, for the photon, the whole universe is contracted into zero length. The photon is emitted and absorbed instantaneously. From the perspective of the photon, it is as if the photon never existed, because what can exist for zero time? What all this means is something way beyond human comprehension, but as Einstein himself said:

> 'The most beautiful experience we can have is the mysterious. It is the fundamental emotion that stands at the cradle of true art and true science. Whoever does not know it and can no longer wonder, no longer marvel, is as good as dead, and his eyes are dimmed.'

CHAPTER 10

What is spacetime?

Einstein showed that even though space and time could be changed, the new concept of spacetime was absolute.

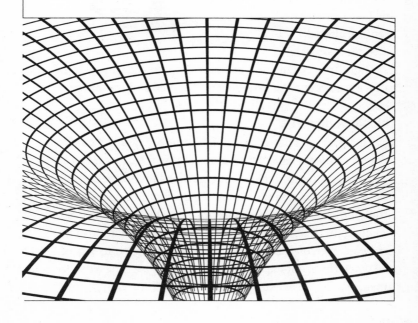

A year after Einstein published his special relativity theory, the German mathematician Hermann Minkowski wrote:

'Henceforth, space by itself, and time by itself, are doomed to fade away into mere shadows, and only a kind of union of the two will preserve an independent reality.'

Einstein argued that absolute time and absolute space could be replaced by absolute spacetime. The mathematical reality of relativity shows that space and time are inextricably linked and both, as we have seen, are altered when we approach near-light speeds. Only by considering space and time together can we accurately describe what is observed at light speed.

Hermann Minkowski

The block universe

To help visualize a path through spacetime, physicists employ a concept called the block universe. Picture the universe as a huge, rectangular box. Now imagine it as a four-dimensional box by adding time as the fourth dimension. Taking a slice through the box gives us a snapshot of the block universe at a moment in time. Any event at any place in our flat universe can be mapped on the box, its coordinates showing us where and when the event happened. In fact, the spacetime block universe maps all events, past, present and future.

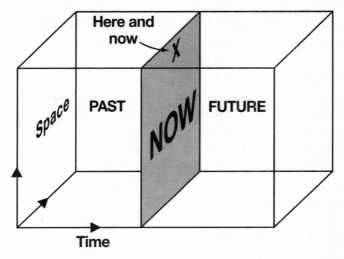

The block universe

How does time flow through the block universe from the present to the past? One view is that 'now' is the slice taken at this very moment. Time 'flows' in a series of infinitesimal jumps from slice to slice, with each jump so small that we could never detect it. Another view has it that all of the slices exist simultaneously, with all past and future events mapped along the timeline, but we are prevented from seeing this as we cannot step outside the four dimensions of spacetime.

> *'The ordinary adult never gives a thought to spacetime problems . . . I, on the contrary, developed so slowly that I did not begin to wonder about space and time until I was an adult. I then delved more deeply into the problem than any other adult or child would have done.'*
>
> Einstein writing in a letter to Nobel laureate James Franck. Einstein believed it was usually children, not adults, who pondered spacetime problems.

Whether time flickers into the future, like the pages of a flipbook, or whether the future already and unalterably exists,

Einstein's relativity ties us into a picture of the universe in which space and time are inextricably linked. As we have discovered, the effects of time dilation and length contraction mean that spacetime divides into its space part and time part differently for observers in reference frames moving relative to one another. There is no way, for example, to claim unambiguously that an event lasted ten seconds without giving some indication of the reference frame in which the measurement was made. Observers in relative motion will not, in other words, be able to agree upon which page of the flipbook an event took place.

Minkowski spacetime diagrams

In 1907, Hermann Minkowski developed another way of visualizing how objects moved through space and time. These representations in spacetime are called Minkowski spacetime diagrams and they allow us to visualize some of the odd effects of relativity.

In a Minkowski spacetime diagram, a coordinate system is used, with time shown vertically on the y axis, and either one or two of the space dimensions represented along the x and z axes, like a perspective drawing. If we were to think of a Minkowski diagram in a similar way to the block universe diagram, then in Minkowski's case the time slices are stacked vertically, with the past at the bottom of the pile. Each of these slices is called a spacelike hypersurface. In reality, these spacetime snapshots are three-dimensional, not flat surfaces, but as we saw with the block universe, visualizing four-dimensional space is hard work!

In a Minkowski diagram, an object is not represented as a single point but as a line containing all the spacetime points at which it exists. This is the object's worldline. If it is in uniform motion, the object's worldline will be straight, but any force acting on it will cause the worldline to curve. If one object's worldline meets that of another object then the two objects collide at that point. The units along the time axis are usually given as seconds × the speed of light, so that the worldlines of light rays make a 45-degree angle with each axis.

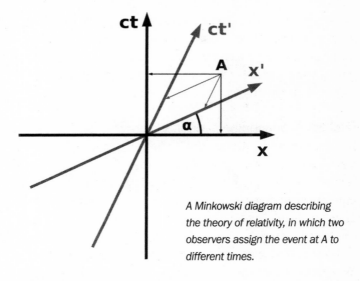

A Minkowski diagram describing the theory of relativity, in which two observers assign the event at A to different times.

The fact that nothing can travel faster than light places a restriction on how events can influence one another in spacetime. The possible light-speed worldlines caused by an event spread out from it in a growing circle, like ripples spreading across a pond from where a fish has leapt up. Imagine all these split-second circles stacking one on top of the other up the timeline. As the circles, each one bigger than the one before, stack along the timeline they form an inverted cone shape, with its point at the origin of the event. This is called a light cone. The future light cone of the event maps out all the possible future events in spacetime that the event can affect. Because nothing can travel faster than light, anything outside the light cone cannot possibly be influenced by, or have any knowledge of the event.

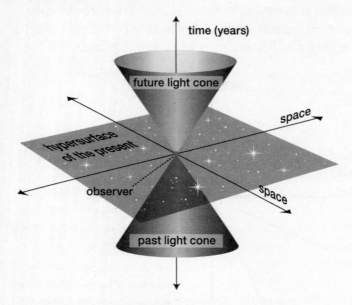

time (years)

future light cone

space

hypersurface of the present

observer

space

past light cone

There is also an exactly symmetrical past light cone expanding out from the event into the past. The past and future light cones divide spacetime into three regions. The absolute future of the event is the region inside the future light cone. It contains everything that can possibly happen as a result of the event. The absolute past of the event is everything inside the past light cone. It contains everything that could possibly have caused or affected the event. Anything outside the past light cone can have had no effect on – or caused – the event in question. Everything that lies outside the past and future light cones of an event is said to be 'elsewhere'. Anything in the 'elsewhere' can have no knowledge of the event and can have no influence on it or be affected by it.

The British physicist Stephen Hawking illustrates the usefulness of light cones by describing a scenario that would take place on Earth if the Sun were suddenly to go out. Because of the time it takes the Sun's light to reach Earth, we would know nothing about the extinguishing event until the Earth entered the Sun's future light cone eight minutes later. Until then, we would be quite unaffected by the fact that the Sun had gone out.

It isn't essential to have light in order to have a light cone. The light cone is simply a map of the geography of spacetime which shows the limits of the possible interactions that can be had with the event. Every event in spacetime has its own light cone, which means that spacetime is filled with an endless number of infinitely overlapping cones.

Spacetime and simultaneity

Spacetime diagrams can be used to explain some of the puzzling effects of special relativity, such as time dilation and length contraction.

Any event that occurs simultaneously with an event on an observer's worldline will lie on a hypersurface that is perpendicular to that worldline. In other words, all points on the hypersurface lie at the same point in time, though they may be widely separated in space.

Now imagine a second observer, moving relative to the first. The second observer's worldline follows a path that is at an angle to the first's, which means that their hypersurface slice will also be tilted with respect to the first. Observer 2 cannot agree with observer 1's notions of which specific events are simultaneous with each other.

Before Einstein's revelations in 1905, it was widely accepted that everyone experienced time in the same way. In fact, we all have a slightly different version of the spacetime flipbook.

Travelling through spacetime

Imagine you're in an aircraft flying due south. The pilot makes a course correction so the aircraft is now flying south-west. The aircraft is still travelling in a southward direction, but not as quickly as before because part of its velocity is now taking it west as well. What does this have to do with spacetime?

In Newtonian physics, travel through time and travel through space were held to be two quite separate things. But, according to Einstein, the two are inextricably linked. If you are stationary – that is, not moving through space – then all your spacetime movement is through time. When you start to move, some of your movement through time is diverted into movement through space. Like the plane changing course, the speed of your journey through time slows down when some of your motion is used for your journey through space.

Because the speed of light is constant, one observer's measurements of space and time differ from those of another observer in relative motion, so that each measures the same value for the speed of light. According to special relativity, the combined speed of an object's motion through time and through space is precisely equal to the speed of light. This is an upper speed limit which can't be broken. For an object in motion, time must slow down otherwise the total combined speed through spacetime would exceed the speed of light. At the speed of light all the spacetime movement has become movement through space with nothing left over for movement through time. Which is why a photon of light, as we saw earlier, travels instantaneously across the universe from its own perspective.

These relativistic effects on time are greater the closer we are to the speed of light, but they apply to any movement through space, even at slow speeds. Experiments carried out using atomic clocks have demonstrated that clocks on an airborne plane run a few hundred billionths of a second slower than clocks that remain on the ground. It is a small difference to be sure, but it matches exactly with the predictions of special relativity.

Absolute spacetime

Einstein was never entirely happy with his theory being called 'relativity' theory. As he had demonstrated, some things were relative, such as motion, distances and the duration of time, but they all took place against the unchanging backdrop of spacetime, the geometry of which was rigidly dictated by the speed of light. Absolute spacetime is as crucial to the understanding of special relativity as were absolute time and absolute space to Newtonian physics. Einstein preferred to think of his work as invariance theory, founded on the unvarying nature of spacetime and the unchanging speed of light. Physicist Abraham Pais, who wrote probably the best scientific biography of Einstein, said there were two things that Einstein was particularly good at: 'He knew how to invent invariance principles and how to make use of statistical fluctuations.'

An invariant is something that stays constant under various transformations. A sphere is invariant, because it looks the same no matter how you turn it round. A cube, however, is invariant only under 90° rotations – if you turn it from face on to edge on, it will look different. Einstein's insight in the special theory was that the speed of light is an invariant. It is constant, no matter who measures it or how fast they are travelling at the time.

Why does E = mc²?

It's the most famous formula in science, but what does it mean?

So far, we've looked at what happens to objects in uniform motion. As we have seen, as an object approaches light speed, time slows down and its length contracts along the direction of travel. But what causes an object to be in motion? As Newton stated, an object will remain at rest or in uniform motion unless acted upon by a force. Einstein's theory changed the study of forces and motion (dynamics) and led to the famous equation $E = mc^2$.

Einstein published his idea in a short paper as a sort of coda to the special theory of relativity. It was entitled 'Does the Inertia of a Body Depend Upon its Energy-Content?' and submitted it to the *Annals of Physics* in September 1905.

Force and momentum

An object in motion is said to have momentum, which is a measure of the quantity of motion. It is defined by the equation:

Momentum = mass × velocity.

Increasing either the mass or the velocity of an object increases its momentum. When two moving objects collide, energy and momentum are transferred between them. This exchange results in a force acting on the two objects. A force is a measure of the rate of transfer of energy and momentum. Force, energy and momentum are related to one another by the following formulae:

Momentum gained = force × time during which the force acts
Energy gained = force × distance through which the force acts

These formulae are true both in classical Newtonian physics and Einstein's relativistic physics.

WHAT IS ENERGY?

The E in Einstein's equation stands for energy, but what is energy? The modern use of the term dates from the 1840s, when it was first used by physicist William Thomson (who later became Lord Kelvin). He realized that the power that drove many different processes could be explained in terms of the transfer of energy from one system and form to another. Energy comes in different forms. For example, there's the chemical energy stored in our muscles that allows us to move, there's kinetic energy, the energy of movement, there's potential energy, such as the energy stored in a tightly drawn bow-string, there's electromagnetic energy, heat energy, and atomic energy.

Without energy, nothing would happen. The more available energy there is, the more that can be achieved. If an object is described as energetic it means it can do things. Scientists believe that the amount of energy in the universe is limited. Energy can change from one form to another but it cannot be created or destroyed.

Lord Kelvin

Just as energy is conserved in any interaction, so too is momentum. Newton's third law of motion

For every action there is an equal and opposite reaction

arises from this. When a rocket is launched, for example, its upward momentum is balanced by the downward momentum of the hot gases expelled from its engines.

In classical physics, it was theoretically possible to impart as much momentum to an object as you wanted and accelerate to any speed you desired. All you had to do was apply a large enough force for a long enough time, and it should be possible to exceed even the speed of light. But this is something that relativity theory simply does not allow.

In relativistic physics, it is also possible to impart unlimited momentum to an object by applying a force to it. But no matter how much or for how long a force is applied, the object will never accelerate beyond the speed of light.

In the classical framework it was naturally assumed that the mass of the object stayed the same and an increase in momentum meant an increase in velocity. But Einstein decreed that this wasn't the case – maintaining that as the velocity of an object increases, so does its mass. As the object approaches light speed, less of the increase in momentum is taken up by an increase in velocity and more is taken up by an increase in mass.

WHAT IS MASS?

Mass is the amount of 'stuff' something contains. You can measure mass by measuring weight, which is the gravitational mass. A 10 kg bag of potatoes is definitely more massive than a 5 kg bag, but a 10 kg sack of potatoes will only weigh about 1.6 kg on the Moon even though you will still have the same amount of potatoes as you had on Earth. In other words, their mass will be the same. Mass is also a measure of the amount of inertia, or resistance to movement, an object has – its inertial mass. From Newton's equation

force = mass × acceleration (F = ma)

you can determine how massive an object is from the amount of force you have to apply to get it moving.

Just like the effects of time dilation and length contraction, the mass increase isn't felt by the object itself. The crew of a spacecraft approaching light speed don't find themselves getting heavier. The increase in mass is only apparent to an external observer who is relatively stationary compared to the spacecraft, and who sees that it is resisting being accelerated.

According to relativity theory, an object cannot be accelerated beyond the speed of light because the closer it gets to the speed of light, the more its mass increases towards infinity. It becomes harder and harder to accelerate it because the amount of energy required to do so also increases. When Einstein published his theory, it was known that it was increasingly difficult to accelerate electrons in a cathode ray tube as they approached light speed. At the time, it was thought this was caused by some relationship between the electron and the electromagnetic field, but Einstein showed it was the result of an increase in the electron's mass.

Kinetic energy

The energy of movement, kinetic energy, is given by the equation:

$$KE = \tfrac{1}{2}mv^2$$

This means that the kinetic energy of an object is equal to half its mass multiplied by the velocity squared. This works well for everyday 'low' speeds, but is increasingly inaccurate approaching light speed as m, the mass, begins to increase.

A moving object increases in mass and has kinetic energy by virtue of its motion. As a moving object slows down, it loses kinetic energy. An object at rest has zero kinetic energy. The mass of an object cannot be zero of course. The lowest mass an object can have is called its rest mass and its mass while in motion is its relativistic mass.

Finally, $E = mc^2$

If an object is moving almost at the speed of light then any force acting on it (and imparting energy and momentum) will cause it to increase in mass because it can't go any faster. We have seen that energy gained equals the force multiplied by the distance through which the force acts. At almost the speed of light, the distance the object moves is nearly the same as the distance travelled by light in the same time. We can write this as an equation:

$E = force \times c$
(where E = energy and c = the speed of light)

As momentum equals mass × velocity and the velocity doesn't change during the time in which the force acts, the mass increases by an amount and the velocity remains at close to the speed of light. We can represent this as:

$Force = m \times c$
(where m = mass)

These two equations can be combined into:

$E = force \times c = (m \times c) \times c$

Which is simplified as:

$E = mc^2$

Two sides of the same coin

Einstein's equation says that energy and mass are the same thing. If an object gains or loses mass or energy, it gains or loses an equivalent amount of energy or mass in accordance with $E = mc^2$. Does it hold true for other forms of energy, as well as kinetic energy? For example, does an object lose mass as it cools down? Yes, it does: temperature is a measurement of the speed at which the atoms and molecules that make up a substance are moving around, so in accordance with $E = mc^2$, the faster they move, the more massive they'll be.

Einstein believed his formula would explain a curious discovery made by the Polish physicist Marie Curie. She had observed that an ounce of radioactive radium produced 4,000 calories of heat per hour, seemingly indefinitely. Where, she wondered, was this energy coming from? According to Einstein, as the radium radiated heat it should also be losing mass. But the equipment available at the time wasn't accurate enough to measure the tiny amount of mass that was being converted into energy and there was no way to verify Einstein's explanation. Einstein wrote: 'The idea is amusing and enticing, but whether the Almighty is laughing at it and leading me up the garden path – that I cannot know.'

Years later, in 1948, Einstein explained the equivalence of mass and energy:

> 'It followed from the special theory of relativity that mass and energy are both but different manifestations of the same. . . . Furthermore, the equation $E = mc^2$, in which energy is put equal to mass, multiplied by the square of the velocity of light, showed that very small amounts of mass may be converted into a very large amount of energy and vice versa. The mass and energy were in fact equivalent, according to the formula mentioned before.'

The speed of light is a very big number; squared, it is a very big number indeed. If it were possible to convert even a tiny amount of matter into energy, the output would be enormous. Physicist Richard Wolfson calculated that there is almost enough energy stored in a raisin to provide all New York City's needs for a day.

CHAPTER 12

How did Einstein fit gravity into relativity?

Special relativity was only the beginning, now Einstein had to find a way of bringing gravity into his calculations.

In setting out the special theory of relativity, Einstein concentrated solely on objects moving with uniform motion. He chose to ignore objects that were accelerating and objects affected by gravity. He did this for the very good reason that it made the calculations a lot easier. Writing in *The Times* newspaper on 28 November 1919, he said:

> '...the theory of relativity resembles a building consisting of two separate storeys, the special theory and the general theory. The special theory, on which the general theory rests, applies to all physical phenomena with the exception of gravitation; the general theory provides the law of gravitation and its relations to the other forces of nature.'

Formulating the general theory would take Einstein seven years of intense work. Physicist Dennis Overbye described his achievement as 'arguably the most prodigious effort of sustained brilliance on the part of one man in the history of physics'. What emerged was a view of the universe totally unlike any that had gone before.

The gravity conundrum

Special relativity was built around the fact that light was the fastest thing in the universe, but this directly contradicted Newton's ideas about how gravity worked. According to Newton, gravity made its effects felt instantaneously – from the Sun holding the Earth in orbit, to the path of a space probe to Pluto or a skydiver falling back to Earth, gravity acted without delay, apparently propagating through space faster than light. If the Sun were to disappear, the Earth would slingshot out of its orbit in the same second – there would be no waiting eight minutes for it to enter the Sun's light cone. But if Einstein was right, then this couldn't be possible.

Newton's universal law of gravity had been backed up by observation and experiment time and time again. So how did gravity make its influence felt? Obviously, it was a force that acted at a distance and didn't require any physical contact to work. Also, unlike any other force, it was impossible to shield yourself from its effects. Newton's law explained how to calculate the effects of gravity, but didn't explain what caused it. Passing the buck somewhat, he wrote in the *Principia*: 'I leave this problem to the consideration of the reader.'

Acceleration

As Einstein had shown, if you are in uniform motion then it's impossible to demonstrate that you are actually moving. Accelerated movement is quite different. If you change speed or direction, you will feel it. Without looking out of the window, you know when a train is rounding a bend because you feel yourself leaning sideways. Similarly, when an aircraft begins to accelerate down the runway for take-off you feel yourself pressed back into your seat. Without any visible clues, you can tell when an elevator has started to move up or down. When we accelerate, we feel inertial forces – the forces that resist a change in speed or direction. These are the forces that throw us against the side of the train when it takes the bend; and they are the forces that cause our coffee to part company with its cup when the bus we are travelling in hits a pothole.

One, two, free fall

In his legendary experiment at the Tower of Pisa, Galileo demonstrated that it will take the same amount of time for a small stone and a large stone to fall to the ground. This is because the two stones accelerate at the same rate and this holds true no matter what the difference in their mass. Galileo couldn't explain why this was the case, but Newton did so with his second law of motion – force = mass × acceleration. The acceleration of a falling object on Earth is always the same – 9.8 mps (other planets differ). This idea, that gravity accelerates all objects at the same rate regardless of what they might be made of, is called the 'Universality of Free Fall' or the 'Equivalence Principle'. Einstein would build his theory of gravity by assuming the Equivalence Principle to be true.

This came about because two quantities in Newtonian theory, the inertial mass of a body and its gravitational mass, matched exactly. Einstein believed that this couldn't be coincidence. If he was going to find a workable theory of gravity it would have to explain this phenomenon.

Einstein's happiest thought

In what he called his 'happiest thought', which he most likely had sometime in November 1907, Einstein realized that gravity and acceleration were equivalent – in other words, without a frame of reference you can't tell one from the other. Lecturing in Kyoto, Japan, in 1922, he said:

> 'I was sitting in a chair in the patent office at Bern when all of a sudden a thought occurred to me: If a person falls freely he will not feel his own weight. I was startled. This simple thought made a deep impression on me. It impelled me towards a theory of gravitation.'

Should you have the extreme misfortune to be in an elevator on the top floor of a skyscraper when the cable snaps, you will immediately begin plunging to your doom. However, you should have time to observe some intriguing phenomena. Your feet are no longer pushing against the floor of the lift. Should you choose to jump off the floor you will not fall back. It is as if gravity has vanished. There are no experiments you can perform that will conclusively show whether you are heading towards the ground or floating freely in deep space far from any gravitational force. If you close your eyes (you probably will), you can imagine yourself a weightless astronaut. The laws of physics are the same for the plummeting lift occupant and the astronaut.

Einstein in a box

In another of his famous thought experiments, Einstein imagined a physicist who awakens in a box. Unknown to the physicist, the box is no longer on Earth but in deep space and under uniform acceleration. If the physicist were to release objects in the box, their inertia would cause them to fall to the 'bottom' of the box, that is, in the opposite direction to that in which the box is moving. All objects dropped by the physicist would fall exactly alike, no matter what their mass or composition, in accordance with the theories of Galileo and Newton. The physicist would therefore conclude that there was a gravitational field at work inside the box.

Einstein asserted that there was indeed a gravitational field. He formulated a principle of equivalence which stated that the effects of uniform acceleration are indistinguishable from the effects of gravity. Acceleration creates a gravitational field. According to Einstein's equivalence principle, the physicist inside the box would consider himself to be in a gravitational field and not accelerating, but an observer would see the box accelerating uniformly through gravity-free space. Each relative point of view is equally valid. This was what made the inertial mass and the gravitational mass the same.

Redshift

Einstein's equivalence principle predicts that the wavelength of electromagnetic radiation will lengthen as it climbs out of a gravity well, a phenomenon called gravitational redshift.

THE VOMIT COMET

NASA trains its astronauts on board a modified aircraft that flies on a parabola-shaped free-fall flight path. This allows the astronauts to experience near weightlessness for around 20 seconds. The plane is known as the 'Weightless Wonder' or the 'Vomit Comet' as the effects of weightlessness can induce nausea. The aircraft can fly 40 to 60 parabolas in the course of a research flight. First, the pilot takes the aircraft up at a 45-degree angle before throttling back the engines to slow it and pitch the nose down to complete the parabola. As the aircraft pulls out of its dive, the passengers and crew feel a force equivalent to twice that of normal gravity.

Colonel John Stapp undergoes a high G-force test.

Thanks to Einstein's $E = mc^2$, and Planck's $E = hf$ law relating the energy of light to its frequency, it becomes apparent that as a photon moves out of a gravitational field, it must lose energy. Since photons always travel at the speed of light, this energy loss is seen as a lowering in frequency rather than as a reduction in velocity. This lowering of the frequency of the photon corresponds to a 'redshift' to the lower-frequency, longer-wavelength end of the spectrum.

Another consequence of redshift is the slowing of time. If we send a beam of light from the surface of the Earth to an observer high above it, he or she will see a lowering of its frequency, which means that the length of time between

Gravitational red-shifting

one wave crest and the next has increased. It appears to the high-flying observer that things down below are taking longer to happen. This prediction of general relativity was tested in 1962 when two extremely accurate atomic clocks were placed in a tower, one at the top and one at the bottom. The clock at the bottom, the one deepest in Earth's gravity well, ran more slowly than the one at the top.

Solving the twins paradox

Earlier, we looked at the paradox of the twin in the spaceship travelling at close to light speed who returns home to discover she is now younger than the sibling who stayed behind. The twin in the spaceship had to accelerate to get close to light speed (and decelerate to slow down, turn around and come back). Acceleration is equivalent to gravity and gravity, as we have seen, slows time. It follows, therefore, that acceleration also slows time. The spacefaring twin ages more slowly than

LIGHT BENDS, TIME SLOWS

Einstein realized that one consequence of the equivalence principle is that the path of a light beam will be bent by gravity. Imagine a photon crossing the physicist's box as it accelerates through space. As the photon is crossing the box, the floor is accelerating upwards, which means that the photon appears to fall downwards.

A second consequence of the equivalence principle is that time slows in a gravitational field. This effect, called gravitational time dilation, means that observers at different distances from a large object (which produces a gravitational field) will obtain different measurements for the time elapsed between two events. This is a direct consequence of the fact that an observer outside the box (outside the gravitational field) sees the photon follow a straight path, but the physicist inside the box sees it follow a longer, curved path. Because the speed of light cannot change, the physicist's clock has to run slower to allow both journeys to be made in the same time.

her sibling because she was accelerating while her sibling was not. Once again, this demonstrates that there is no absolute time in relativity; time is something personal to everyone, measured according to where we are and how we are moving.

To the equivalence principle, Einstein added the relativity principle, which stated that the laws of physics in any frame of reference are governed by the theory of special relativity. These were the foundations upon which he built his theory of general relativity, which extended the concept of spacetime to all of physics and, particularly, to the theory of gravitation.

CHAPTER 13

How does Einstein define gravity?

For centuries, gravity was believed to be a force of attraction between two objects, but to Einstein it was a distortion in the very fabric of spacetime.

What effect does gravity have upon spacetime? The key idea in general relativity is that it is not a force acting between objects, but the result of a distortion of spacetime caused by the objects in it. The bigger the object, the more spacetime curves around it.

Gravity and tidal forces

Imagine two spacecraft moving along parallel trajectories through empty space at the same speed relative to each other. Provided that no force acts on them, their paths will follow a straight line. Now suppose there's a planet up ahead. According to Newton, its gravity exerts a force that will tug the spacecraft off course, causing their paths to converge. This happens because they are both being pulled towards the planet's centre of gravity, so they are both moving towards the same point in space. It is this difference in direction that's responsible for the distance decreasing between the two spacecraft. Physicists call this force difference a 'tidal force', because it is just such a difference between the Moon's gravitational attraction on Earth and the Earth's oceans which causes the tides to rise and fall.

Tidal forces also show us that gravity is not entirely 'switched off', even in free fall. For a human-sized object free-falling in Earth's gravity, there will be a tidal force acting because the object's feet, closest to the ground, are feeling a very slightly stronger gravitational pull than the object's head, which is further away from the centre of gravity.

It's an insignificantly small difference, but it is there nevertheless.

Curved surfaces

We are familiar with the geometry of flat plane surfaces. The sum of all the angles in a triangle is always 180 degrees; two parallel lines will never meet, and so on. But there *is* a way in which two parallel paths can converge.

The surface of a sphere is also a flat surface. There are no straight lines on a sphere, or on any type of curved surface, but we can construct lines that are as straight as possible. Mathematicians call these straight-as-possible lines 'geodesics'. In the case of the sphere, the shortest distance between two points will lie along the path of a great circle, the largest circles that can be drawn on that surface. Great circle routes are the ones taken by commercial airline pilots to ensure that they follow the shortest flight path between two airports.

Instead of spacecraft, now imagine two supersonic skimmers heading north in parallel from the equator to the North Pole. Without changing course, they will collide at the North Pole when their paths meet. No force has acted on them, but because they are moving across the curved surface of a sphere their courses will inevitably cross.

To see why this happens we can construct a triangle on the surface of the sphere from three intersecting geodesics. The bottom of the triangle lies on the equator and represents the distance between the two skimmers. The two angles on the equator are both right angles, so the initial flight paths are parallel. But the paths converge at the North Pole tip of the triangle. The angles of the triangle are going to add up to more than 180 degrees.

It is only when you look at larger regions of the sphere that it becomes apparent the surface is curved. You could walk all day and not get any impression that you were moving along the surface of a sphere, but if you were to rise a dozen kilometres or so into the air, you would just be able to discern the curvature of the Earth. The same is true for any curved surface: a small enough section of the surface is effectively indistinguishable from a flat plane.

Curved spacetime

Einstein took this property of curvature and drew a comparison with the way gravity works. For a very small region of spacetime – for example, a physicist floating freely in a box in space moving uniformly – there is no gravity. The inside of the box obeys the laws of the spacetime of special relativity, where gravity is absent. This is analogous to the flat surface and the laws governing motion are fairly straightforward. As long as no force acts on an object, it will continue to move in a straight line at a constant velocity, following a straight path through spacetime.

If we add gravity to the situation, by placing a large planet in the path of the physicist, for example, Newton dictates that the planet will exert a force on all objects around it. The hapless physicist will begin to feel the first effects of acceleration as his path begins to curve towards the planet.

In his theory of gravity, Einstein looks at things in a different way. Rather than exerting a force, a mass causes a distortion of spacetime. Empty spacetime, the spacetime of special relativity, is flat. But where there is matter present, spacetime is curved. In the same way that there are no straight lines on the surface of a sphere, there are no straight lines in curved spacetime. The closest we can get to the straight line in curved spacetime, just as on a sphere, is a geodesic, a curve that is as straight as possible. The physicist heading towards the planet has not been deflected from his straight-line course, but the presence of the large planet distorting spacetime has changed the form a straight line can take. It has redefined the geometry of spacetime. According to general relativity, an object follows a straight-line geodesic through spacetime, but from our three-dimensional perspective the path looks curved.

The Earth makes a dent in spacetime, curving it around itself. The Moon follows a straight path through the Earth-curved spacetime, which to us appears to take it on a circular orbit (though it's elliptical, in fact, as the Moon curves spacetime as well) around the Earth. General relativity predicts that light rays will be bent by gravitational fields

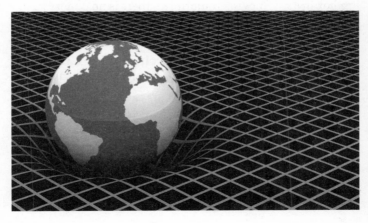

because light also follows geodesics through spacetime. This bending of light by gravity was, as we shall see, one of the first confirmations that Einstein's theory was correct.

A dance to the music of spacetime

This is the basis of Einstein's theory. Newton's gravity is a force that acts on objects and influences their movement, but gravity in Einstein's universe is the result of curved spacetime, a distortion of spacetime geometry. Objects still follow the straightest possible paths through spacetime, but because spacetime is now curved they accelerate as if they were under the influence of a gravitational force.

In Einstein's universe, matter and spacetime interact in a complex and ever-changing dance. Matter distorts the geometry of spacetime and this distorted geometry dictates how matter moves through it. As the matter moves and the sources of gravity change positions, so the swirling curves of spacetime ebb and flow. As the physicist John Archibald Wheeler summarized it:

> 'Spacetime tells matter how to move; matter tells spacetime how to curve.'

Gravitational waves

One of the predictions of general relativity was that there should be a phenomenon called 'gravitational waves'. These are like ripples in spacetime which are caused by particularly energetic disturbances.

Einstein's equations showed that cataclysmic events, such as two black holes colliding or a massive supernova explosion, would be like large rocks being dropped into the pond of spacetime, sending out waves of distorted space across the universe at the speed of light.

Although gravitational waves were predicted to exist in 1916, there was no proof of their existence until 20 years after Einstein's death. In 1974, astronomers at the Arecibo Radio Observatory in Puerto Rico discovered a binary pulsar – two extremely dense and heavy stars in orbit around each other. Knowing that this system could be used to test Einstein's prediction, the astronomers began making careful observations of the system. Eight years of meticulous data-gathering revealed that the pulsars were getting closer to each other at exactly the rate predicted by general relativity. After over 40 years of close monitoring, the observed changes in the orbits of the pulsars is in such close agreement with general relativity that the researchers have no doubt that the system is emitting gravitational waves.

Until September 2015, all the confirmations of the existence of gravitational waves had been indirect or determined mathematically, not through physical proof. On 14 September, the Laser Interferometer Gravitational-Wave Observatory (LIGO) in the USA detected gravitational waves for the first time. They were generated by two colliding black holes nearly 1.3 billion light years away. They may be generated by extremely violent events, but by the time the waves reach the Earth they are many millions of times smaller. By the time the gravitational waves detected by LIGO had reached the Earth, the degree of spacetime wobbling they generated was much smaller than the nucleus of an atom.

LIGO is a triumph of engineering skill and ingenuity. It consists of two L-shaped detectors built 3,000 km apart and houses vacuum chambers 4 km long. It can measure a motion 10,000 times smaller than an atomic nucleus – nothing of this accuracy has ever been attempted before. It is equivalent to measuring the distance to the nearest star to a precision smaller than the width of a human hair.

CHAPTER 14

How did an eclipse prove Einstein was right?

Arthur Eddington's astronomical observations during a total eclipse of the Sun confirmed that Einstein's relativity equations were correct.

In the autumn of 1919, Pauline Einstein received a postcard from her son. It began:

'Dear Mother, joyous news today. H. A. Lorentz telegraphed that the English expeditions have actually demonstrated the deflection of light from the Sun.'

When Einstein first set out his equivalence principle in 1907 and deduced that it would result in the bending of light, he thought the effect would be far too small ever to be measured. His first predictions for the bending of light from a star were in accord with what Newton himself would have predicted from his law of gravitation and belief that light took the form of a stream of particles. But Einstein's solution was incorrect.

At this point, he had not yet discovered that spacetime was curved and that this would have an effect on the bending of the light beam. It was not until 1915 that, according to his general relativity theory, Einstein realized that the light ray passing the Sun would be bent by twice the value of that initial 1907 calculation. It is perhaps fortunate that there had been no opportunity to test Einstein's idea before he made this correction. In 1912, an expedition to view an eclipse in Brazil included measuring the deflection of light in its list of experiments, but bad weather prevented this from going ahead. In the summer of 1914, a second expedition set off for the Crimea to observe an eclipse, but it was forced to turn back when World War I broke out.

Now, with a clear difference between Einstein's predictions from general relativity and those of Newtonian physics, it should be possible to find out who was right, but further observations had to be put on hold until the end of the war. Some attempts were made to find evidence of the predicted deflections in photographs of earlier eclipses, but without success. Einstein was anxious to have his theory proved right. In 1916, he wrote a book in an attempt to explain relativity to a wider audience, stating:

> 'The examination of the correctness or otherwise of this deduction is a problem of the greatest importance, the early solution of which is to be expected of astronomers.'

Then, in September 1919, two British expeditions finally obtained the results he had been hoping for.

The eclipse expeditions of 1919

The English astronomer Sir Arthur Eddington led an expedition to the island of Principe, off the coast of West Africa, during the total eclipse of 29 May 1919. A second expedition, to Sobral in Brazil, was led by Andrew Crommelin from the Royal Observatory in London.

Eddington had been lucky enough to obtain a copy of Einstein's theory in 1916. He became an enthusiastic champion of relativity and, together with the Astronomer Royal, Sir Frank Dyson, came up with a plan to test Einstein's theory.

The stars are only visible at night because, during the day, their faint light is dimmed by the radiance of the Sun. During a total eclipse, when the Moon blocks the light from the Sun, the stars become visible. Einstein's theory predicted that the path of a star's light, passing close to the Sun on its way to Earth, would be deflected by the warped spacetime around the Sun. This would result in a change in the

Sir Arthur Eddington

apparent position of the star which could be measured against its actual position, known from observations of the star at night. The angle of deflection that Eddington and Dyson were looking for was very small indeed, roughly equivalent to the diameter of a coin seen from a distance of 3 km, but it was achievable, even with the technology available at the time.

The two expeditions set out from Liverpool in March 1919, one group heading to Brazil and the other to Principe. On the morning of the eclipse it was raining heavily on Principe and the required observation seemed hopeless. During the morning the sky began to clear but, as totality (total eclipse) approached, the last few tatters of clouds still threatened to frustrate the observations. Eddington reported that he didn't see the eclipse beyond a couple of glances because he was too busy changing the photographic plates in his camera. He was concerned that the cloud would prove to have interfered with the star images.

The Sobral team were luckier with the weather. In order to discover whose theory was correct – Newton's or Einstein's – the plates from both expeditions would have to be developed and carefully examined. One of the Brazil photographs appeared to agree with Einstein, but another agreed with Newton. Eddington's plates showed fewer stars, but what was visible seemed to support Einstein's thinking. Eddington decided that the Newtonian result from Brazil had been produced by faulty equipment and he declared Einstein vindicated.

Einstein was meeting with one of his graduate students, Ilse Schneider, shortly after he had heard the news from Eddington. Schneider asked him what he would have done if the observations had shown his theory to be wrong. Einstein famously replied, 'Then I would have been sorry for the dear Lord; the theory is correct.'

Space warped

In the UK, on 7 November 1919, *The Times* newspaper ran some momentous headlines. 'Armistice and treaty terms', 'Devastated France' and 'War crimes against Serbia' featured on page 11. On page 12 there was more news concerning the aftermath of World War I, including the 'Armistice day observance/Two minutes' pause from work'. But if the reader were to scan across to column six, he or she would find some world-changing news of a different kind: 'Revolution in science/New theory of the universe/Newtonian ideas overthrown'. And there, halfway down the column,

Einstein with Arthur Eddington in 1930

a subheading that would have resulted in much scratching of heads: 'Space warped'.

In the same newspaper, on 28 November, Einstein wrote:

> 'By an application of the theory of relativity to the tastes of readers, today in Germany I am called a German man of science and in England I am represented as a Swiss Jew. If I come to be regarded as a bête noire, the descriptions will be reversed and I shall become a Swiss Jew for the Germans and a German man of science for the English!'

His words were strangely prescient given the events that would take place in Germany during the following decades.

A message from Mercury

Astronomy played another role in demonstrating the validity of the general theory. A long-standing puzzle for astronomers was the fact that the orbit of Mercury, the planet closest to the Sun, did not quite fit in with Newton's equations.

As the planets orbit the Sun, they follow an elliptical path; this had been determined by Johannes Kepler in 1609 and was explained by Newton some 50 years later. The elliptical path means that the planet has a point of closest approach to the Sun; this is called the 'perihelion' (the farthest point from the Sun in the planet's orbit is called the 'aphelion'). The perihelion doesn't always occur in the same place on every orbit, but slowly moves around the Sun. The reason for this is because of the pull of the planets on one another (an effect predicted by Newton). This change in the rotation of the planet's orbit is called 'precession'.

Newton could explain the precession of all of the planets except that of Mercury, whose rate was a bit greater than he predicted. It was a small difference, but it couldn't be ignored.

On Christmas Eve 1911, Einstein wrote:

'At this time I am [again] busy with considerations on relativity theory in connection with the law of gravitation. . . . I hope to clear up the so-far unexplained changes of the perihelion length of Mercury . . . [but] so far it does not seem to work.'

Astronomers hunted for a way to explain Mercury's odd behaviour. Perhaps there was a swarm of asteroids between Mercury and the Sun, or perhaps even an undiscovered planet was tugging on Mercury as it orbited. There were many suggestions, but none seemed to answer all the questions. What they all had in common, however, was that they accepted Newton's law of gravitation as accurate.

In 1916, Einstein pitched in with the equations from his newly forged general theory of relativity. He demonstrated that his concept of how gravity worked exactly predicted Mercury's orbital movements. The reason for the discrepancy was the warping of spacetime in close proximity to the huge mass of the Sun. It appeared that Einstein's theory was right and his calculations agreed with the astronomers' observations. 'For a few days, I was beside myself with joyous excitement,' he wrote.

CHAPTER 15

If Einstein was right, was Newton wrong?

The nature of scientific enquiry is that it never stands still, no truth is absolute, and no proof immune from being overturned.

Isaac Newton may have been wrong in the tiniest details but for 200 years or so he was undoubtedly right in practice. It is the very nature of scientific enquiry that truths are never absolute, but simply the best description of reality that can be achieved with the knowledge available at the time. A law of science is an explanation or a statement that always appears to be true. Newton's law of gravity and his three laws of motion are excellent at explaining why objects move the way they do.

'Newton, forgive me; you found the only way which, in your age, was just about possible for a man of highest thought and creative power. The concepts which you created are guiding our thinking in physics even today, although we now know that they will have to be replaced by others farther removed from the sphere of immediate experience, if we aim at a profounder understanding of relationships.'

Albert Einstein

But in 1905 Einstein came along and showed that for objects travelling at rates approaching the speed of light, Newton's laws didn't apply any longer. Newton wasn't wrong – he just couldn't imagine or foresee the limits of his laws.

Newton offered a way of calculating the effects of gravity, but he never sought to explain them. In 1687, he wrote:

'I have not yet been able to discover the cause of these properties of gravity from phenomena and I frame no hypotheses. . . . That one body may act upon another at a distance through a vacuum without the mediation of anything else, by and through which their action and force may be conveyed from one another, is to me so great an absurdity that, I believe, no man who has in philosophic matters a competent faculty of thinking could ever fall into it.'

Einstein's explanation for the phenomenon of gravity happened to provide much more accurate predictions than

William Blake's famous painting of Sir Isaac Newton

Newton's, whose laws still work well in 'normal' (that is, at well below the speed of light) conditions. They're certainly accurate enough to plot a course to send a probe from the Earth to Pluto. Einstein didn't prove Newton was wrong; he produced a theory of gravity that extended to cover situations of which Newton had no knowledge.

It is one of the guiding principles of science that a theory is nothing if it can't stand up to experimental scrutiny. As Richard Feynman said:

> *'It doesn't matter how beautiful your theory is, it doesn't matter how smart you are, it doesn't matter what your name is. If it disagrees with experiment, it's wrong.'*

A replica of Isaac Newton's telescope

It takes a great deal of experiment to prove a theory right or wrong. One experiment isn't enough because results have to be replicated and verified. Newton's laws stood up to experiment time and time again for over a hundred years. For example, 19th-century astronomers noticed that Sirius, the brightest star in the night sky, appeared to wobble slightly in its path. Newton's laws say that if something isn't moving as expected, a force must be being applied to it. Some might have put forward the alternative argument that this demonstrated how Newton's laws only hold true in our solar system and not in interstellar space. But if Newton was right, then perhaps there was an unseen star orbiting Sirius whose gravity was causing it to wobble. In 1862, this star was discovered. In fact, Sirius is a binary star system – two stars orbiting a point between them. One is a main sequence star, Sirius A, and the other a white dwarf star, Sirius B. Hence the wobble – a vindication for the laws of Newton.

One of the first cracks in Newtonian physics was manifested by his unexplained deviations in the orbit of Mercury. There was nothing in Newton's *Principia* to explain these. Although astronomers searched for a new planet to make sense of Mercury's gravitational pattern, even calling it Vulcan, there was no intrasolar version of Sirius's companion star, Sirius B, to explain it.

Provided that the gravitational fields involved are weak, the predictions of Einstein's general theory of relativity are the same as those of Newton's theory of gravitation. In other words, the two theories concur as long as the velocities of all objects interacting with one another gravitationally are small compared with the speed of light.

A gravitational field is considered strong if the escape velocity required to break free of it approaches the speed of light. All gravitational fields encountered in the solar system, even the one in the vicinity of the Sun, are weak by this definition. At low speeds and in weak gravitational fields, general and special relativity's predictions agree with everyday experience and Newtonian physics.

Einstein had huge respect for Newton's achievements. He wrote:

> 'In one person he combined the experimenter, the theorist, the mechanic, and, not least, the artist in exposition. . . . He stands before us strong, certain, and alone: his joy in creation and his minute precision are evident in every word and every figure.'

On this particular point Newton was wrong but, as far as anyone could know, he was right in his time. Similarly, Einstein is right for now, but one day he too may be shown to be wrong if some yet deeper truth is discovered. As we shall see, even general relativity is not the full story. At the extremes of the cosmos, in places such as black holes where the escape velocity not only approaches but exceeds the speed of light, Einstein begins to fail us. General relativity also breaks down on the subatomic scales where we enter the quantum realm.

ESCAPE VELOCITY

The idea of an escape velocity was one which arose from a study of Newton's laws. Ignoring complicating factors such as air resistance, the escape velocity is the speed an object would need to attain in order to escape a planet's gravitational pull and continue on into space. For example, the escape velocity from Earth is about 11.2 km/s at the surface. In 1783, when it became known that light had a finite speed, physicist John Michell posed an interesting question. Could we not in theory have a very small, but very massive object, with an escape velocity so high that light would not be able to escape from it? If this were so, said Michell, then the most massive objects in the universe might well be dark. This may be the first recorded reference to the phenomenon now known as a 'black hole'.

Like Newton before him, Einstein took a giant step along the path to a fuller understanding of the workings of the universe. His theory of general relativity appears to reflect the way in which nature seems to work, and tests of the general theory's equality between gravitational and inertial mass have been found to be accurate to within one part in 10 trillion, which is as accurate as we can be with current equipment. There are no cracks observable in general relativity into which we might try to slip a new theory.

So general relativity does the job of explaining the universe. But, 200 years ago, the same was said about Newton's law of gravity. As French philosopher Claude Lévi-Strauss put it:

'The scientist is not a person who gives the right answers, he's one who asks the right questions.'

CHAPTER 16

Why didn't Einstein's theory win the Nobel Prize?

Today it seems inconceivable that special relativity theory did not receive world recognition at the time. Why was this?

When Einstein published his ground-breaking papers on the nature of space, light, motion and the atomic realm in 1905 he was a 26-year-old patent clerk, little known outside his immediate circle. He was disappointed by the lukewarm reaction to his ideas.

In 1908, Max Planck also wrote a paper on special relativity. It was largely the result of Planck's importance, as opposed to the insignificance of a technical expert third class in the Bern patent office, that the idea of relativity began to be accepted. Hermann Minkowski also published an important paper on relativity in 1908, and showed that the Newtonian theory of gravitation was not consistent with relativity.

Hendrik Lorentz, who had attempted to explain the result of the Michelson–Morley experiment by suggesting that objects contracted as they moved through the ether, never seemed to accept Einstein's conclusions, even though they echoed some elements of his own thinking. He gave a lecture in 1913 in which he remarked that he found:

> 'A certain satisfaction in the older interpretation according to which . . . space and time can be sharply separated. . . . Finally it should be noted that the daring assertion that one can never observe velocities larger than the velocity of light contains a hypothetical restriction of what is accessible to us, a restriction which cannot be accepted without some reservation.'

Despite their apparent differences, Lorentz and Einstein were jointly proposed for the 1912 Nobel Prize for physics for their work on special relativity. The man who put forward their names was the German physicist Wilhelm Wein, who had won the prize the previous year for his 'discoveries regarding the laws governing the radiation of heat'.

The Swedish Academy and Nobel Museum in Stockholm

Science and politics

Nominations for the Nobel Prize are considered by a five-man (it would have been exclusively men in those days) committee nominated by the Swedish Academy of Sciences. They may be received from previous Nobel laureates (anyone can nominate for any field, not just their own) and from professors in select universities, which in the early 20th century meant, almost exclusively, universities in Nordic- and German-speaking countries. Between 1911 and 1921, Einstein was nominated repeatedly. But two members of the committee, who were chosen to write reports on Einstein's suitability for the prize, repeatedly recommended he should not receive it.

One of them was the 1903 chemistry prize winner, Svante Arrhenius, who was a founder of physical chemistry. A physicist himself, Arrhenius was impressed by Einstein's work on Brownian motion and even thought it was worthy of a Nobel Prize. Yet he argued that it would look strange to award the prize for Brownian motion since Einstein's other work surpassed it; on the other hand, he didn't feel able to nominate him for the later work because he regarded it as still experimentally unproven.

The other member of the committee was Allvar Gullstrand, winner of the 1911 prize for medicine, and he was most vehemently opposed to awarding the Nobel Prize to Einstein. Gullstrand's specialism was the optics of the eye and his interests were in theoretical optics. Many of Einstein's critics simply failed to understand general relativity, but this wasn't true of Gullstrand. Like Arrhenius, he argued that there was little in the way of experimental evidence for special relativity. He remarked to a friend at one point that Einstein 'must never receive the Nobel Prize, even if the entire world demands it'.

Gullstrand, like Arrhenius, argued that there was little empirical evidence in favour of special relativity. General relativity had its three famous tests, but one of these, the gravitational redshift of the Sun, was considered unfavourable to Einstein by most experts until 1922. The 1917 Report of the Nobel Prize Committee refers approvingly to Einstein's

Allvar Gullstrand

work, but also mentions the fact that measurements carried out at the Mount Wilson Observatory in California had not found the redshift which general relativity predicted. 'It appears that Einstein's relativity theory, whatever its merits in other respects may be, does not deserve a Nobel Prize,' the committee concluded. That the redshift is a real phenomenon was confirmed by laboratory experiments at Harvard University in the 1960s.

Another test of general relativity, the bending of light by the Sun, was widely contested, despite the results of Sir Arthur Eddington's 1919 eclipse expedition. The single greatest demonstrable success of relativity theory, either special or general, was Einstein's explanation of anomalies in the orbit of Mercury which couldn't be explained by Newtonian mechanics. Einstein had shown that his theory predicted a perihelion shift, a change in the point in Mercury's orbit where it came closest to the Sun, which agreed precisely with the observed effect. Gullstrand claimed that Einstein had fudged the calculation to fit the result, and said that Einstein's theory could match any result one wished for a given problem.

While attempting to demonstrate the absurdity of relativity, Gullstrand accidentally stumbled upon an important consequence of it: the viewpoint of an observer falling into a black hole. Gullstrand's conception of a black hole was one in which, once inside the event horizon, space was drawn in towards the singularity faster than light (see pages 152–4).

After Eddington's eclipse results became known, nominations for Einstein to be awarded the 1920 prize flooded in, but they were not well received by the committee. According to science historian Robert Friedman, the committee did not want a 'political and intellectual radical, who – it was said – did not conduct experiments, crowned as the pinnacle of physics.' Instead, the 1920 Nobel Prize was awarded to the Swiss physicist Charles-Edouard Guillaume for his discovery of anomalies in nickel steel alloys. When the announcement was made, the previously unknown Guillaume was, according to Friedman, 'as surprised as the rest of the world'.

Einstein the celebrity

In early 1920, following Eddington's confirmation of gravitational light bending, Einstein became something of a reluctant celebrity, sought out for his opinions on all manner of subjects. He was always charming and patient in his dealings with people, but was more comfortable being left to pursue his work. Few understood the nature of relativity but everyone, it seemed, wanted to talk about it.

Einstein with contemporaries in 1931; Albert Michelson is standing next to him, on the left of the photo.

In the 1920s, the journalist Alexander Moszkowski, a German-Jewish satirist, published a book of conversations with Einstein in which he commented on the public passion for relativity:

'In all nooks and corners, social evenings of instruction sprang up, and wandering universities appeared with errant professors that led people out of the three-dimensional misery of daily life into the more hospitable Elysian fields of four-dimensionality.'

Max Born was horrified that Einstein had agreed to collaborate on the book, fearing it would stoke the anti-Semitic sentiment against Einstein that was already making itself known. A growing number of German nationalist scientists had taken to referring to Einstein's ideas as 'Jewish physics'. Einstein himself maintained an air of detachment. 'The whole affair is a matter of indifference to me,' he said, 'as is all the

POWER IN THE ATOM

In 1920, Einstein gave a lecture in Prague, after which a reception was laid on for him by the university physics department. Following a number of enthusiastic speeches, Einstein was invited to respond. Instead of giving the expected address, Einstein announced: 'It will perhaps be pleasanter and more understandable if instead of making a speech I play a piece for you on the violin.' He then went on to perform a sonata by Mozart in what his friend Philipp Frank described as a 'moving manner'.

The next day, according to Frank, a young man approached Einstein in Frank's office. On the basis of $E = mc^2$, insisted the man, it would be possible 'to use the energy contained within the atom for the production of frightening explosives'. Einstein brushed off the suggestion, declaring it foolish.

commotion, and opinion of each and every human being. I will live through all that is in store for me like an unconcerned spectator.'

Winning the prize

Although Einstein was again nominated in 1921, Gullstrand again convinced the committee that none of the year's nominations met the criteria as outlined in the will of Alfred Nobel. According to the Nobel Foundation's statutes, the Nobel Prize can, in such cases, be held back until the following year, and this is what happened.

The 1921 deferral meant that two prizes could be awarded in 1922. Einstein received many nominations for relativity, but was also nominated for his work on the photoelectric effect. This nomination came from Carl Wilhelm Oseen, a Swedish theoretical physicist. Oseen wanted the committee to recognize the photoelectric effect as a fundamental law of nature, not just as a theory. He did so not so much to support Einstein as to champion the work of Niels Bohr, who had proposed a new quantum theory of the atom. This, according to Oseen, was 'the most beautiful of all the beautiful' ideas in recent theoretical physics.

In his report to the committee, Oseen, exaggerating the close ties between Einstein's photoelectric effect and Bohr's new description of the atom, succeeded in his aim. The committee was won over and, on 10 November 1922, awarded the 1922 prize to Bohr and the delayed 1921 prize to Einstein. The citation for Einstein's Nobel Prize reads:

'The Nobel Prize in Physics 1921 was awarded to Albert Einstein "for his services to Theoretical Physics, and especially for his discovery of the law of the photoelectric effect".'

Einstein's Nobel Prize certificate

Einstein was en route to Japan when he heard the news. He did not attend the official ceremony and didn't receive his prize until the following year. He had promised that the monetary

Einstein in Japan in 1922

award from the Nobel Prize would be put in trust for his sons, with his ex-wife, Mileva Maric, allowed to draw from the interest. The prize money was duly transferred to Maric.

CHAPTER 17

What was Einstein's greatest blunder?

We tend to remember Einstein as a scientific genius, but what was his greatest mistake?

In the early part of the 20th century, when Einstein published his general theory of relativity, most people thought the Milky Way galaxy was the whole universe and there was nothing beyond it. Evidence was only just beginning to accumulate to support the notion that the universe was much, much bigger than anyone had previously imagined, and people began to debate whether or not some objects in space might lie outside the Milky Way altogether.

In 1923, Edwin Hubble settled the argument when he used the Hooker Telescope at the Mount Wilson Observatory in California, the most powerful telescope in existence, to study stars in the Andromeda nebula. He estimated their distance at 800,000 light years (an underestimate of some 1.2 million, as it turned out).

Andromeda was a galaxy in its own right, distinct from our own Milky Way galaxy. Hubble went on to find other even more distant galaxies. A picture began to emerge of a universe vast beyond imagining, stretching over billions of light years, with a hundred billion galaxies each containing around a hundred billion stars.

Edwin Hubble

The universe was firmly believed to be static – no one thought it might expand or contract. In 1929, Hubble became famous overnight when he made another momentous discovery. He found that the light reaching us from the distant galaxies is shifted towards the red end of the electromagnetic spectrum (see box overleaf), indicating that these galaxies are moving away from our solar system. The further away they are, the faster they are retreating. The galaxies that are twice as far away move approximately twice as fast. The best explanation for this phenomenon was that the universe must be expanding.

The cosmological constant

Einstein's general theory certainly allowed for the notion that the universe could be either expanding or contracting. In fact, applied to the universe as a whole and not just to a star or planet within it, the equations of general relativity demand that the size of the universe is changing. General relativity doesn't allow for a static universe – it could not exist because the curving of spacetime by the matter in it would eventually cause the universe to collapse in on itself. Therefore, if the universe was neither static nor collapsing, then it had to be expanding. But Einstein, in common with everyone else at the time, thought this notion far-fetched.

If the universe was expanding then, logically, it had to be expanding from somewhere. At some distant point in time, the universe must have started out as a single point containing all space and time. Einstein thought this a nonsensical idea and in 1917 introduced a term called the 'cosmological constant' (he referred to it as a 'slight modification') into his equations. This repulsive force counterbalanced the attraction of gravity and prevented the expansion or contraction of the universe. Einstein was not entirely happy with this addition,

DOPPLER REDSHIFT

The redshift Hubble saw in the distant galaxies wasn't gravitational; it was caused by the Doppler effect. We are familiar with this from the effect of sound waves. If a police car hurtles past, siren wailing, the pitch of the siren gets lower as it travels further away from you. This is because successive sound waves take longer to reach you, and this sounds like a lowering in pitch. When the car is approaching, the opposite occurs – successive sound waves reach you faster and the pitch is higher. A similar effect applies to light being emitted by moving objects, but instead of a pitch change the wavelength of the light is shifted – towards the red end of the spectrum (redshift) if the object is moving away from you, and towards the blue end (blueshift) if it is moving towards you.

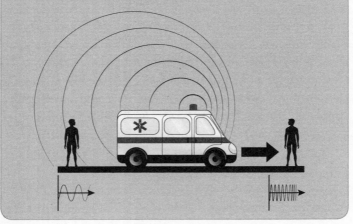

admitting that it was 'not justified by our actual knowledge of gravitation'.

Edwin Hubble's discovery of the redshifting of distant galaxies could not be disputed. He had demonstrated that the universe really was expanding and this was reported in the popular press as a challenge to Einstein's theories. Einstein was happy to admit his mistake and remove the cosmological constant from his equations. He described the astronomers at Mount Wilson as 'outstanding', writing to his friend Michele Besso that the situation was 'truly exciting'. It's a pity that Einstein hadn't trusted his original equations – had he done so, he would have predicted the expanding universe a decade before Hubble confirmed it.

Arthur Eddington and others pointed out that the cosmological constant would not have worked in any event as it required the universe to be in a highly delicate state of balance, somewhat like balancing a pencil on its point. This would mean that the slightest disturbance would trigger a runaway expansion or contraction.

Why doesn't the cosmos collapse?

Some of the problems faced by the astronomers and physicists of the early 20th century had been discussed over 200 years earlier. In 1692, Newton received a letter from the Reverend Richard Bentley. Supposing the universe were infinite, said Bentley; if that were the case, then every part of the universe should feel the pull of gravity, and surely, therefore, it should collapse in on itself?

Newton tried to explain this by arguing that if the stars were evenly distributed in space then the force of gravity would act equally in all directions and a balance would be maintained. He quickly realized that this could not be the case, as the slightest movement of any star would upset the balance and the whole cosmic edifice would come crashing down.

Newton and Bentley had made one major error – the stars are not stationary. (It was partly the notion of the 'fixed stars' that enabled Newton to come up with his ideas about absolute space.) It was Edmund Halley, of comet fame and vindicator of Newton's laws, who first observed that a few stars had shifted from the positions recorded for them on Greek star maps.

Olbers' paradox – why isn't the sky full of stars?

Halley pointed out another problem. If the universe was infinite, then wherever you look there ought to be a star – the entire sky should be shining as brightly as the Sun! It clearly wasn't, and this observation had led Johannes Kepler, in the 17th century, to conclude that the universe couldn't be infinitely large. The problem became known as Olbers' paradox, after the German astronomer Heinrich Olbers (1758–1840).

Olbers suggested that there must be clouds of dust between the stars which hid some of them from our view. But this solution was also flawed. Given sufficient time, the energy from the distant stars would heat up the gas clouds until they glowed and the sky was filled with light.

The answer to the problem was revealed with Edwin Hubble's discovery that the universe is expanding. The light from the furthest reaches of the universe has not had time to reach us – and perhaps it never will. The universe is dark because it started with a bang.

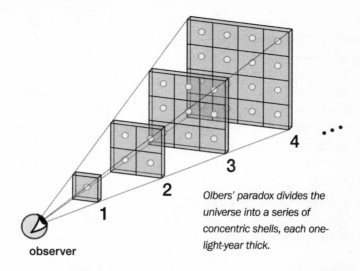

Olbers' paradox divides the universe into a series of concentric shells, each one-light-year thick.

observer

Einstein's biggest blunder?

The physicist George Gamow reported that Einstein called the cosmological constant his 'biggest blunder', but there's no hard evidence that he ever did, and it certainly never appeared in any of his writings. But there was one mistake that Einstein really regretted. After visiting Einstein in Princeton, on 16 November 1954, Linus Pauling wrote in his diary: 'He said that he had made one great mistake – when he signed the letter to President Roosevelt recommending that atom bombs be made.'

The return of the cosmological constant

In the late 1990s, cosmologists made a startling discovery – the universe isn't just expanding, it's expanding at an increasingly rapid rate. The cause of this accelerating expansion is a mystery – scientists refer to a 'dark energy' at work. Most observations support the idea that this dark energy behaves like Einstein's 'cosmological constant' and many cosmologists are keen to revive the term. One speculation is that pairs of so-called 'virtual' particles and antiparticles pop in and out of existence in empty space, a phenomenon allowed for by quantum mechanics. The energy carried by these particles might exert a repulsive force that pushes everything in the universe outwards. The cosmological constant, far from being a blunder, might cause scientists to re-evaluate what they believe to be true about cosmology, particle physics and the fundamental forces of nature.

EINSTEIN'S ENVELOPE

On his second trip to the USA, in 1931, Einstein visited the Mount Wilson Observatory in California with Edwin Hubble. They were met by the frail and elderly Albert Michelson, of the Michelson–Morley ether experiment. As the telescope was being demonstrated, Einstein's wife, Elsa, was informed that it had been used to determine the size and shape of the cosmos. 'Well,' she responded, 'my husband does that on the back of an old envelope.'

CHAPTER 18

Where does Einstein's relativity theory break down?

What happens to relativity at the most extreme limits of reality, inside the event horizon of a black hole?

General relativity has been an incredibly successful means of providing new insights into our understanding of the universe and the way it works. But just like Newton before him, there are some things that Einstein cannot explain.

The more compact and massive the object, the stronger its gravitational influence. General relativity predicts the existence of black holes, regions where the density of matter is so intense that spacetime is warped and curved to the extent that it becomes infinite. The gravity well formed by a black hole is so deep that nothing can break free from it. It is like a hole in spacetime and not even light has a spacetime path it can follow to escape it.

Pulsars and neutron stars

In 1967, Jocelyn Bell, a graduate student at Cambridge University, noticed something unusual in her radio telescope observations. There seemed to be a rapidly pulsing signal coming from a point in the sky. She looked further, and found another regularly pulsing signal. Some people suggested that these might be a sign of intelligent alien life and they were dubbed LGMs (little green men).

The American astronomer Thomas Gold wondered whether they were, in fact, neutron stars. Astronomers had speculated for some time about their existence. Once their nuclear fuel is exhausted, stars that are between five and 40 times as massive as our Sun end their lives in a gigantic outpouring of matter and energy called a supernova. The outer layers of the star are blasted out into space, leading to such a dramatic increase in the star's brightness that it can outshine the other stars in its galaxy for a short time.

At the same time as the outer layers are expanding into space, the star's core collapses. Within seconds, the core's density increases so enormously that the electrons and protons which make it up are squeezed together to form neutrons. The core region becomes an unbelievably dense ball of nuclear matter, where trillions of tons of material are squeezed into each cubic centimetre. A neutron star is no more than 20 km in diameter, but with a mass greater than that of our Sun. If you could bring a teaspoonful of neutron star material to Earth, it would weigh more than a mountain.

As the star collapses it spins faster and faster, like a figure skater on the ice spinning faster by pulling her arms and legs closer to her body. The contracting neutron star may eventually be rotating a few hundred times every second. The star's magnetic field becomes more concentrated and more powerful. Electrons within the magnetic field are accelerated to close to the speed of light and emit beams of electromagnetic radiation from the star's north and south poles. The star is now acting like a cosmic lighthouse, with two narrow beams of electromagnetic waves pointing in opposite directions. We can only detect the neutron star if one of these beams happens to sweep across the surface of the Earth and be picked up by radio telescopes. Because such stars appear to pulse as the beam briefly points towards the Earth, they were given the name pulsars.

Neutron stars and pulsars in particular are ideal cosmic laboratories for testing general relativity. The gravity of a compact, massive body such as a neutron star is very strong and the effects of general relativity therefore become much more apparent. For example, binary systems which have a neutron star orbiting an ordinary star can be used to make precise measurements of the influence of gravity on light.

Black holes

What would happen if a neutron star continued to collapse? In 1928, the Indian astronomer Subrahmanyan Chandrasekhar calculated that, if a star was bigger than a certain size, the power of its gravitational force would be greater than its atomic particles could resist. The star would simply continue collapsing until it formed a single point, a singularity, warping spacetime around it to such an extent that nothing could escape, not even light. It would become a black hole in space.

A black hole isn't a physical object, it is more a region of spacetime with very peculiar properties. The border that separates this region from the rest of the universe is called the event horizon. The event horizon is a one-way door – matter or energy can cross over it from the outside, but can never return. Therefore, an observer will detect no light from a black hole. This means it is impossible to observe one directly, but it is possible to see the effect a black hole has on its surroundings.

The light of stars and galaxies passing by a black hole is bent by gravitational influence, much as the Sun's gravity bends the starlight that passes it but to a much greater degree. If an object and a black hole are precisely aligned, an observer would see the light from the object bend to form a ring around the black hole. This phenomenon is called an Einstein ring. If the star is out of alignment – even if only slightly – the ring can't be seen and the black hole is much more difficult to detect.

An Einstein ring

Astronomers need to look for indirect ways to deduce the presence of a black hole, for example, by looking for unexpected movements in nearby stars. If a black hole forms from a star that was part of a binary system, it may start to pull gases in towards itself from the outer layers of its neighbouring star. This gas swirls in around the black hole, forming an accretion disc that reaches such high temperatures that it emits X-rays. A neutron star can have a similar effect, but sometimes observations suggest that the object is too compact to be a neutron star. In that event, astronomers think they have a good case for saying they've found a black hole. To the best of our knowledge, the nearest black hole is more than a thousand light years distant.

WHAT IS A SINGULARITY?

One consequence of Einstein's description of gravity in terms of curves in spacetime is that it allows for the formation of singularities. A singularity is a point at which some property is infinite. For example, the density of the material at the centre of a black hole is infinite because the mass of the star has been compressed into zero volume under the effect of infinite gravity. At the centre of a black hole, spacetime has infinite curvature and space and time ceases to exist in any meaningful sense. The laws of physics – including those of relativity – break down in the singularity.

Singularities were anathema to Einstein, who believed there was no place for such infinities in a proper mathematical description of the universe. He argued that singularities could not appear in nature 'for the reason that matter cannot be concentrated arbitrarily . . . because otherwise the constituting particles would reach the velocity of light.'

How big is a black hole?

A black hole can be any size. Black holes that form when a large star becomes a supernova have a radius of around 5 km. Galactic black holes, which form in the cores of some galaxies, can have a mass equal to that of millions of stars and be bigger than our solar system. At the other end of the scale, mini black holes, which formed in the early universe, may be smaller than a grain of sand but as massive as mountains. The outer boundary of a black hole – its event horizon – forms at a point called the Schwarzschild radius. This is the radius below which the gravitational attraction between the particles making up an object becomes so intense that it will suffer an irreversible gravitational collapse, eventually forming a black hole. Theoretically anything can form a black hole if you squash it down hard enough. The Schwarzschild radius for an average human being is about 10–23 cm – smaller than the nucleus of an atom.

The Schwarzschild radius was discovered by the German astronomer Karl Schwarzschild in 1916 while he was studying Einstein's general relativity equations. Schwarzschild used it to describe how spacetime would be warped near a spherical star. At the time, Schwarzschild was unable to present his findings to the Prussian Academy because he was busy calculating trajectories for the German army's artillery on the Russian front, so he sent his work to Einstein, who presented it on his behalf.

As a consequence of the severe warping of spacetime, strange effects occur at the event horizon of a black hole. An observer watching someone fall towards the event horizon – if such a thing were possible – would see their clocks running slower and slower until, at the event horizon itself, time appeared to freeze. For the person falling, the opposite would be true. He or she would see time speed up in the rest of the

SUPER-MASSIVE BLACK HOLES

Not long after the invention of the radio telescope, astronomers were finding evidence of highly energetic radio galaxies. Like a neutron star, these shoot out beams of highly energetic particles in opposite directions, but on a vastly greater scale. When these beams interact with clouds of intergalactic gas, they cause them to emit radio waves which can be detected by telescopes on Earth. It was clear that there was only one possible power source – matter falling towards a compact mass and forming a high-energy accretion disc. It was on a scale so enormous that the central mass would have to be extremely massive and extremely compact.

Astronomers now believe these are super-massive black holes with more than a million times the mass of the Sun, and that they are found in the core of galaxies, even relatively peaceful ones such as our own Milky Way. The current super-massive heavyweight champion weighs in at a mass of 21 billion Suns, which is pretty much incomprehensible on any human scale of understanding. It can be found in the crowded Coma galaxy cluster, which consists of more than 1,000 galaxies.

universe, and perhaps even witness its end before crossing the event horizon.

Einstein did not believe that black holes could form, but other theorists did. They showed how a sufficiently massive star would inevitably collapse at the end of its life to form a super-massive singularity where all the laws of physics, including Einstein's, would break down.

CHAPTER 19

How did relativity lead to a Big Bang?

Relativity suggested that the universe hadn't existed forever, but had begun at some point.

After Einstein introduced the world to general relativity, a number of scientists, Einstein included, tried to see how the theory would apply to the universe as a whole. At the time this required them to make assumptions about how matter in the universe was distributed. Their first notion was that if you look on a large enough scale the universe appears more or less the same whichever direction you look at it. Their second thought was that the universe looks the same wherever you are in it; in other words, the matter in the universe is homogeneous (the same everywhere) and isotropic (the same in every direction) when averaged over very large scales. These assumptions constitute the Cosmological Principle.

Armed with Einstein's description of how gravity works and an idea of how the matter is distributed in the universe, it is possible to start building a picture of how the universe has evolved over time. The picture we get is one in which the universe started out, effectively, from zero. If we take the expanding universe and throw it into reverse, rewinding its evolution back in time, we see all matter, all energy, all space and all time contract down into a single point of infinite density and gravity and zero size – a singularity, in other words. At this point, Einstein's equations, the foundation stones of his general theory that describe how the distortions of spacetime affect the matter and energy embedded in it, break down – just as they do in a black hole singularity. For whatever reason, and we just don't know why, everything that makes up the universe today expanded from this zero point in an event that came to be known as the Big Bang (see illustration on the facing page).

How did it all begin?

The first person to suggest a possible beginning for the expanding universe was the Belgian priest and astronomer

Georges Henri Lemaître (1894–1966). He put forward his idea of a 'primordial atom' at the end of the 1920s in a famous paper entitled, 'A Homogeneous Universe of Constant Mass and Increasing Radius Accounting for the Radial Velocity of Extra-Galactic Nebulae'.

Lemaître started with a solution to Einstein's equations corresponding to an expanding universe. From this, Lemaître deduced that the speed of the far galaxies is proportional to their distance – a finding which agreed with Hubble's redshift calculations (see page 144). He suggested that, in the distant past, all the mass in the universe had been concentrated into a single super-atom. According to Lemaître, this primordial atom began dividing again and again, eventually giving rise to all the matter we see today. He didn't use the expression 'Big Bang', but referred to it as a 'day without a yesterday'.

Lemaître met Einstein for the first time in October 1927, during the Fifth Solvay Congress of Physics in Brussels. Einstein had read Lemaître's paper, and they went on to discuss it. Einstein could not fault Lemaître's mathematics, but disagreed with his interpretation, going so far as to call it 'abominable'. At this point Einstein still clung to the idea of a static universe and his Cosmological Constant.

When they met again in 1933, Einstein had abandoned this theory and was more receptive to Lemaître's idea. He had accepted the idea of an expanding universe but not the notion of an initial singularity. He suggested to Lemaître that he modify his model, hoping that if certain changes were made the initial singularity could be avoided. Lemaître soon showed that the revised model also led to a singularity.

A cooling universe

Following World War II, Ralph Alpher and George Gamow, his PhD supervisor, suggested that the universe had been formed from a hot soup of atomic particles at a temperature of trillions of degrees. They called this atomic soup 'ylem'.

The ylem cooled as it expanded because its energy spread over greater and greater volumes of space. Gamow and Alpher wanted to prove that the initial conditions they had imagined were the source of the elements in the universe. Some elements, such as hydrogen and helium, had been found to be very common, whereas others, such as tin and gold, were much rarer.

They spent several months on their calculations, working out the effects of the fall in temperature and density as the universe expanded. Their results suggested that hydrogen and helium should be by far the most common elements and that there should be ten atoms of hydrogen for every one atom of helium. This was exactly the ratio that had been determined by astronomers.

In 1948, Alpher, this time with Robert Herman, published a paper in which they predicted that radiation from the early beginnings of the universe should still be detectable. They calculated that this 'cosmic background radiation', as it was called, would have a temperature of about minus 268°C. This was the last faint glow left over from the unimaginable burst of energy which had given birth to the universe. Alpher tried to persuade astronomers to search for these echoes from the beginning of time. Unfortunately, the equipment didn't exist

Ralph Alpher

to refute or verify their theory, so the prediction was more or less forgotten for nearly 20 years.

In 1964, US radio astronomers Arno Penzias and Robert Wilson made a discovery that finally came down in favour of the Big Bang. While testing an astronomical microwave detector called the Holmdel Horn Antenna, they were concerned to discover that the device seemed to be picking up noise from all over the sky. At first, they thought pigeon droppings might be causing it to malfunction. But after cleaning out the detector – and having the pigeons shot! – they found that the noise was coming from outside the atmosphere, and from every direction. It never varied, whatever time of day they ran the test.

Around this time, physicists Bob Dicke and Jim Peebles were planning an experiment to test Alpher's idea that there should still be traces of radiation from the early universe. They anticipated that the light from the earliest stars would have been so massively redshifted on its epic journey across the universe, that it would appear to us as microwave radiation.

A map of cosmic background radiation, leftover heat from the Big Bang

Hearing of their work, Penzias and Wilson contacted Dicke and Peebles to ask if their own findings were what Dicke and Peebles had been looking for. Dicke confirmed that the mysterious signals were indeed the cosmic background radiation and therefore proof of the Big Bang. 'We've been scooped', he admitted.

WHAT'S IN A NAME?

Alpher had to contend with many astronomers who refused to accept that the universe had a beginning. In 1950, British astronomer Fred Hoyle gave a radio talk on the subject and mocked the ideas of Alpher and Gamow, referring to their theory as a 'big bang'. A fierce opponent of the notion of an expanding universe, Hoyle preferred his own 'steady state' theory in which the universe remains much as it always has been. But the term 'Big Bang' lodged in the public imagination and from that point on, the idea that the universe started from an initial point was known as 'Big Bang' theory.

Where do we go from here?

Since we've established that we are in an expanding universe, what happens next? The fate of the universe depends on the balance between the rate of expansion, which is expressed by a factor called the Hubble Constant, and the curving of spacetime by gravity, which is determined by the amount of matter in the universe.

There are three possible outcomes. In the first scenario, the amount of matter in the universe is greater than the 'critical density', as cosmologists call it, and the expansion is slowed down, stopped and reversed by gravity. Spacetime bends in on itself like a four-dimensional cosmic beachball and the

> 'When we first heard that inexplicable hum, we didn't understand its significance, and we never dreamed it would be connected to the origins of the universe. It wasn't until we exhausted every possible explanation for the sound's origin that we realized we had stumbled upon something big.'
>
> Arno Penzias

Robert Wilson (left) and Arno Penzias

universe eventually collapses again in a Big Crunch. In the second, the density of the universe is a little less than the critical density. The universe continues to expand, but at a slower and slower rate. In the third scenario, the rate of expansion gets faster. The final possible outcome appears to be what is happening. The gulfs between galaxies, already huge beyond imagining, grow ever wider.

Results from the WMAP space probe and observations of distant supernovae seem to indicate that the expansion of the universe is accelerating. This implies the existence of an unknown force acting counter to gravity, which is sometimes referred to as 'dark energy'. The existence of this force is reminiscent of Einstein's cosmological constant.

The amount of matter in the universe also determines its geometry. If the density of the universe is greater than the critical density, the geometry of space is closed and curved like the surface of a sphere. If the density of the universe is less than the critical density, then the geometry of space is open (infinite) and curved like the surface of a saddle. If the density of the universe exactly equals the critical density, then the geometry of the universe is flat, like a sheet of paper, and infinite in extent.

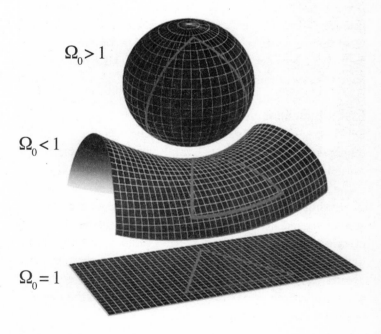

$\Omega_0 > 1$

$\Omega_0 < 1$

$\Omega_0 = 1$

Current thinking predicts that the density of the universe is very close to the critical density, and that therefore the geometry of the universe is flat. In case you're wondering what the critical density is, it corresponds to roughly six hydrogen atoms per cubic metre – not really an awful lot to hang the fate of a universe on!

Faster than light

Believe it or not, some galaxies are moving away from each other faster than the speed of light. How is this possible? Was Einstein wrong after all?

The important thing to remember is that the galaxies aren't all rushing pell-mell through space. It's the universe that's expanding and space itself that's getting bigger, and it is carrying the galaxies along with it. So, although it's impossible to move through space faster than the speed of light, the rule doesn't apply to space itself and it is indeed possible for the distances between galaxies to increase faster than the speed of light.

Could it, should it?

According to general relativity, the beginning of the universe could have been in a Big Bang. But the question that British physicist Roger Penrose posed was: Does general relativity predict there should have been a Big Bang? In other words, is saying that something could have happened the same as saying that it did?

In 1965, Penrose combined the way general relativity explains the behaviour of light cones with the fact that gravity is always an attractive force. He demonstrated mathematically that a star collapsing under its own gravity will eventually become trapped in a region of space that shrinks to zero. Contained within that zero volume, the density of matter and the curvature of spacetime are infinite. The collapsed star forms a black hole singularity.

At the same time that Penrose was working out his theorem, Stephen Hawking was looking for a subject for his PhD thesis. He read Penrose's work and realized that by reversing the direction of time in the theorem (scientifically, a perfectly valid thing to do), and therefore instead of collapsing to zero,

Sir Roger Penrose

expanding out from zero, the theorem still held. Penrose had shown that a collapsing star must end in a singularity. Hawking demonstrated that if the current model of an expanding universe was correct, then it had to have begun with a singularity.

In 1970, Penrose and Hawking produced a joint paper which offered mathematical proof that if the description of the universe given by Einstein's general theory of relativity was correct, and the universe contains as much material as we observe it does, then the universe must have begun with a singularity.

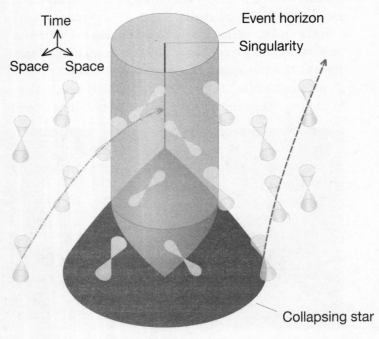

The action of light cones near a black hole

Written years before the work of Hawking and Penrose, Einstein's last words on the Big Bang were:

> '*One may . . . not assume the validity of the equations for very high density of field and matter, and one may not conclude that the "beginning of expansion" must mean a singularity in the mathematical sense.*'

So far, Einstein's general theory of relativity has stood the test of time and experiment. There has been nothing to make scientists doubt its validity as a means of explaining the universe as it is now. But it is an incomplete picture. The theory can't describe what happened at the start of the universe, because it predicts the breakdown of all physical laws, including itself, in the singularity. There must have been a time in the very early history of the universe when events were dominated by the rules of that other great pillar of modern science – quantum mechanics.

Does God play dice?

Another theory – the strange science of quantum mechanics – would send shockwaves through the world of physics.

For the past century, physics has been dominated by two great theories about how the universe works. The practically simultaneous arrival of Max Planck's quantum theory in 1900 and Albert Einstein's relativity theory in 1905 marked the beginning of a period in which the very foundations of physics would be rebuilt.

Einstein's general relativity theory works on a grand scale, describing how gravity shapes the universe of space and time. Quantum mechanics describes how the universe works on the very smallest scale, down to the size of atoms and even smaller. The quantum realm is often described as an 'Alice in Wonderland' world, where events are mysterious, uncertain and inexplicable.

In Lewis Carroll's Alice in Wonderland, *a young girl enters a dream world in which the everyday suddenly becomes weird, and familiar social and scientific rules no longer apply.*

Both theories have been tested by observation and experiment to extraordinary levels of accuracy and each one seems to reflect the universe as it really is. The problem facing physics is that the two theories simply don't join up. The laws of relativity that govern the universe on the large scale don't apply on the small scale of quantum mechanics. The opposite is also true – quantum mechanics tells us nothing about the movements of galaxies or the geometry of the universe. At the moment, there is no theory that successfully combines gravity with quantum mechanics.

Bohr's model

In 1913, the Danish physicist Niels Bohr came up with a theory for the structure of the atom that was based on Planck's and Einstein's ideas about quanta. He wanted to explain how atoms could emit light quanta and also why it was that electrons didn't spiral down into the nucleus when they lost energy. To this end, he theorized that the electrons in an atom remain at fixed distances from the atomic nucleus, arranged in orbits, or shells, around the atom.

With his model, Bohr explained how electrons could jump from one orbit to another by emitting or absorbing energy in fixed quanta. For example, if an electron jumps one orbit closer to the nucleus, it must emit energy equal to the difference of the energies of the two orbits. Conversely, to jump to a higher orbit the electron must absorb a quantum of light equal in energy to the difference in orbits.

Bohr's theory worked well in describing the single electron of the hydrogen atom, but ran into difficulties when applied to larger atoms with multiple electrons. Its insistence on a limited set of permitted orbits seemed somewhat arbitrary, and it looked like a bit of a dead end. The new theory of quantum mechanics would resolve the difficulty.

PLANCK AND EINSTEIN

Max Planck and Albert Einstein had huge respect and affection for each other. On the occasion of Planck's 60th birthday, Einstein spoke of his friend's 'inexhaustible persistence and patience' as he devoted himself to 'the most general problems of our science without letting himself be deflected by goals which are more profitable and easier to achieve. I have often heard that colleagues would like to attribute this attitude to exceptional will-power and discipline; I believe entirely wrongly so. The emotional state which enables such achievements is similar to that of the religious person or the person in love; the daily pursuit does not originate from a design or programme but from a direct need.' The same could probably have been said about Einstein himself.

Wave–particle duality again

In his description of the photoelectric effect, Einstein had shown that light has both wave-like and particle-like properties. In an experiment carried out in 1922, the American physicist Arthur Holly Compton highlighted the dual-wave-and-particle nature of electromagnetic radiation.

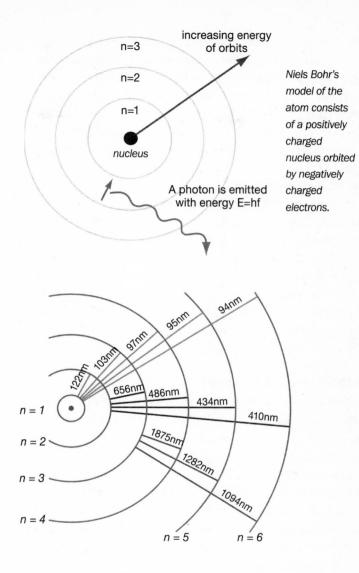

When an electron jumps down to a lower orbital, a photon of a specific wavelength is emitted producing the characteristic spectrum of the atom.

The experiment involved sending a beam of X-rays through a target material. Compton observed that a small part of the beam was deflected off to the sides at various angles and the scattered X-rays had longer wavelengths than the original beam. This change could only be explained by assuming that the X-rays were particles with discrete amounts of energy and momentum, and applying the conservation of energy and conservation of momentum laws to the collision between the photon and the electron.

When X-rays are scattered, their momentum is partially transferred to the electrons with which they interact. The electron takes some energy from an X-ray quantum, and as a result the frequency of the X-ray is reduced. The change in momentum and the frequency shift caused by the scattering are both explained by Einstein's quantum formula.

Compton's experiment had settled the existence of photons, which until then had been disputed. Quantum mechanics followed within months of this revelation.

Is there an electron there?

In the 1920s, researchers were studying the way in which a beam of electrons bounces off a piece of nickel. In this experiment, the nickel crystals acted in a similar fashion to the two slits used in the light interference experiment. Effectively this was the same experiment, but using a stream of electrons rather than a beam of light. The experiments found that, just as the light did, the electrons formed an interference pattern. They were acting as if they were waves, as de Broglie (see page 180) had predicted. But what was the nature of these waves?

The German physicist Max Born said the wave was like a graph mapping out the probability of finding the electron in a particular place. The electron is most likely to be found where

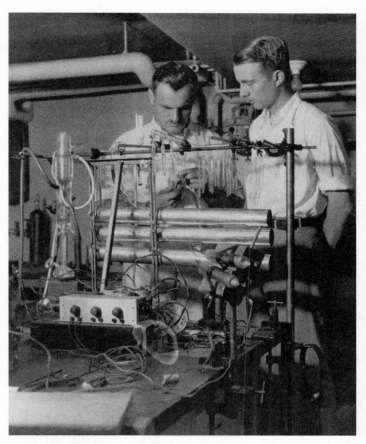

Arthur Compton (on the right) with Luis Alvarez in 1933

the magnitude of the probability wave is large; it is less likely to be found if the magnitude of the probability wave is small. This is one of the stranger notions in quantum mechanics. How can a particle be perhaps here, or perhaps there?

According to Born, the wave nature of matter means that everything has to be examined in terms of probabilities. There is nothing hard and fast in the quantum realm.

WAVE–PARTICLE, PARTICLE–WAVE?

In his 1932 doctoral thesis, the French physicist Louis-Victor de Broglie proposed that not just light but all matter and radiation have both particle- and wave-like characteristics. Appealing to an intuitive belief in the symmetry of nature and Einstein's quantum theory of light, de Broglie asked: if a wave can behave like a particle then why can't a particle, such as the electron, behave like a wave? De Broglie reasoned that as Einstein's famous $E = mc^2$ related mass to energy and Einstein and Planck had related energy to the frequency of waves, then combining the two would suggest that mass should have a wave-like form as well. Einstein supported de Broglie's idea because it seemed a natural continuation of his own theories. Asked by the PhD board to comment on de Broglie's thesis, he said:

'I believe de Broglie's hypothesis is the first feeble ray of light on this worst of our physical enigmas. It may look crazy but it is really sound!'

Louis-Victor de Broglie

The best we can ever do is to say that the electron is likely to be somewhere, we can never say with certainty that it is there. This means that we can carry out an experiment involving electrons and not get the same answer every time, even though we do everything in exactly the same way. All we can measure are probable outcomes, not certain ones.

Heisenberg's uncertainty principle

Werner Heisenberg's famous uncertainty principle, first formulated in 1927, showed it is impossible to know both the position and the momentum of a particle accurately at the same time. The more accurately the particle's momentum is measured the less accurately its position can be determined. If it were possible to measure an electron's momentum with absolute precision, its location would become completely uncertain – we might know how fast it was moving but we wouldn't have a clue where it was.

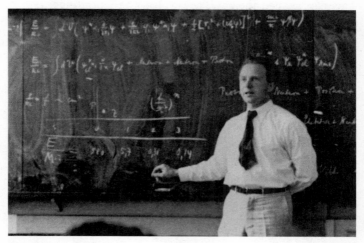

Werner Heisenberg

If the classical physicists had been surprised by wave–particle duality then the uncertainty principle must have left them reeling. We cannot know that such a thing is possible from everyday experience. If you are driving your car, for example, you have a reasonable idea of its position and velocity. But Heisenberg's uncertainty principle asserts that you may know you are travelling at 70 mph, but you have no idea if you are heading for Guildford or Glasgow.

These uncertainties have nothing to do with a lack of skill in the observer or with inadequate equipment. Heisenberg showed that the uncertainty in the momentum multiplied by the uncertainty in the position of the particle can never be smaller than Planck's constant – it is a fundamental property of the universe that puts a limit on what we can know.

Schrödinger's waves

In 1926, the Austrian physicist Erwin Schrödinger worked out an equation which determines how these probability waves are shaped and how they evolve. The Schrödinger equation describes the form of the probability waves (or 'wavefunctions') that govern the motion of small particles. It also specifies how these waves are altered by external influences. Schrödinger tried out his equation on the hydrogen atom, and found it predicted its properties with great accuracy.

It has been said that the Schrödinger equation is as important to quantum mechanics as Newton's laws of motion were for classical mechanics. Schrödinger was trying

Erwin Schrödinger

to describe the quantum world in mathematical terms; he wasn't trying to build up a model you could picture in your head, like the old idea of an atom as a mini solar system. Quantum mechanics was showing how the realm of the atom could be described in precise and rigorous mathematical terms, but with outcomes that could only be seen in terms of probabilities, not certainties.

According to quantum mechanics, when we perform a measurement to locate the position of a particle, we cause its probability wave to collapse. It's no longer possible for it to be somewhere else when you know where it is – the probability of it being elsewhere drops to zero, while the probability of it being where you observed it rises to 100 per cent.

There is still disagreement as to whether or not a wavefunction is a real physical thing or just a mathematical tool that allows us to calculate quantum realm probabilities, but with no basis in reality. From a practical perspective, it doesn't seem to matter. The Copenhagen interpretation of quantum theory (see page 184), developed in the 1920s principally by physicists Niels Bohr and Werner Heisenberg, treats the wavefunction as no more than a tool for predicting the results of observations. It says that physicists shouldn't try to imagine what 'reality' looks like – an approach that physicist David Merman memorably summed up as 'Shut up and calculate!' The certainties of the clockwork universe of Newton's day had been left a long way behind.

The Copenhagen interpretation

In the Copenhagen interpretation of quantum mechanics, championed by Niels Bohr and others (and named after the city where Bohr was based), the properties of a quantum particle have no definite value until a measurement is made. The principle of complementarity is central to the Copenhagen interpretation. This dictates that the wave and particle nature of objects are complementary aspects of a single reality, like the two sides of a coin. An electron or a photon, for example, can behave sometimes as a wave and sometimes as a particle, but never both at the same time, just as a tossed coin can be either heads or tails, but not both simultaneously.

Niels Bohr said it was meaningless to ask what an electron really is. Experiments designed to measure waves will see

Niels Bohr

THE SOLVAY CONFERENCES

Instituted by the Belgian industrialist Ernest Solvay, the Solvay Conferences on physics and chemistry are held in Brussels. The first physics conference took place in 1911 and the first chemistry conference in 1922. They keep to a three-year schedule, with a physics conference in the first year, no conference in the second year, and a chemistry conference in the third year. The fifth physics conference in 1927 is famous for the part it played in formulating the ideas that would come to dominate quantum mechanics and for the dispute between Albert Einstein on the one hand and Niels Bohr, Werner Heisenberg and Max Born on the other, regarding the validity of the Copenhagen interpretation.

The world's most eminent physicists met to discuss the newly formulated quantum theory at the Solvay Conference of 1927. Einstein is pictured in the middle of the front row.

waves, while experiments designed to measure particle properties will see particles. It is impossible to design an experiment that would allow us to see wave and particle at the same time. The wavefunction is a complete description of a wave/particle. When a measurement of the wave/particle is made, its wavefunction collapses. Any information that cannot be obtained from the wavefunction does not exist.

In the 1920s, Max Born stated that particle waves are measures of probability. They're not physical entities like sound or water waves. We can never be certain how any given particle will behave – identical electrons might do different things every time an experiment is run, so the outcome can only be predicted statistically.

The Copenhagen interpretation of quantum mechanics opened a sharp divide between classical Newtonian physics and quantum physics. In the world of everyday things, we expect each event to have a cause. The glass of water didn't fall by itself, it was knocked over when a person stumbled and bumped into the table upon which it was resting. We might not have predicted that someone was going to do that, but if we'd had all the circumstantial details, such as the person's stride length, and the height of the wrinkle in the rug he or she tripped over, we could have come up with a reasonable probability of the spillage occurring. So, according to classical physics, all the variables are there, even if sometimes they are difficult to measure.

In the quantum world, however, there are no relevant details to take into account, there is just pure chance. The quantum world is pure statistical probability. The Copenhagen view is that indeterminacy is a fundamental feature of nature and not just a result of our lack of knowledge. We just have to accept that this is how things are and not attempt to explain it.

Einstein speaks out

Some physicists, among them Albert Einstein, worried about this lack of explanation. In the spring of 1927, the 200th anniversary of Newton's death, two decades after Einstein had with seeming ease overturned much of classical physics with his theory of special relativity, he came to the defence of classical mechanics and causality. 'The last word has not been said,' Einstein argued. 'May the spirit of Newton's method give us the power to restore union between physical reality and the profoundest characteristic on Newton's teaching – strict causality.'

Einstein never came to terms with quantum theory; he believed that it was correct as far as it went, but was fundamentally incomplete. He couldn't accept a reality defined by uncertainty, probability and indeterminacy, believing that the probabilities of quantum mechanics resulted from a gap in our knowledge about how the universe operates on the atomic scale. Once we had a fuller understanding, Einstein thought, probability would be replaced by certainty.

Einstein once said to a friend that when he was judging a theory he asked himself whether, if he were God, 'I would have arranged the world in such a way'. He couldn't believe that there were rules governing most of what happened in the universe but at the fundamental quantum level of reality things seemed to be left to chance. In a letter to Max Born, written in 1926, he stated:

'Quantum mechanics is certainly imposing. But an inner voice tells me that it is not yet the real thing. The theory says a lot, but does not really bring us any closer to the secret of the "old one". I, at any rate, am convinced that He does not throw dice.'

EINSTEIN AND THE LASER

Among all his achievements, it is perhaps not so widely known that Einstein also contributed to the development of the laser. 'Laser' is an acronym for 'Light Amplification by Stimulated Emission of Radiation'. It is a device that creates and amplifies a narrow, focused beam of light, and traces its roots to Einstein's 1917 paper on the quantum theory of radiation.

In a laser, atoms or molecules of either a crystal, such as ruby or garnet, or a gas or liquid are 'pumped' to push them up to higher energy levels. This produces a burst of light as the atoms discharge a flood of photons. This is called stimulated emission, a process Einstein first suggested as a possibility in his 1917 paper. After completing work on the general relativity theory the previous year, he had turned to exploring the interplay of matter and radiation. It was in the course of this work that he came up with an improved fundamental statistical theory of heat, which included the quantum of energy.

Einstein proposed that an excited atom, one that has absorbed a photon, can return to a lower energy state by re-emitting the photon, a process he called spontaneous emission. He also predicted that as light passes through a substance, it could stimulate the emission of more light. His idea was that if you had a large number of atoms in an excited state, all ready to emit a photon, a stray photon passing by could stimulate them to release their photons. These released photons would have the same frequency and direction as the original photon. A cascade of photons is released as the identical photons move through the rest of the atoms. Einstein never put his theory into practice – this would have to wait until 1960, when the first working laser was built.

CHAPTER 21

Who won the argument?

Einstein and Bohr debated quantum theory back and forth for decades. Could either man be said to have won the debate?

Over a period of three decades, until Einstein's death, Einstein and Bohr repeatedly challenged each other's beliefs and interpretations of the quantum world. These debates were never acrimonious – the two physicists were great friends – but each man held to his point of view and defended it stubbornly. Einstein believed there was an objective reality which existed and could be measured, whereas Bohr believed that the very act of measurement altered reality. For example, an electron has no definite position in space until someone decides to measure it.

The Einstein–Bohr contest

Einstein and Bohr crossed paths a number of times during the course of the fifth Solvay Conference on physics. Einstein understood what the quantum mechanics were proposing but felt it wasn't the complete picture. Bohr, who had been confident that Einstein would agree with the Copenhagen interpretation, was shocked and dismayed by Einstein's opposition.

After the conference, Einstein and Bohr entered into a series of contests in which Einstein would attempt to find flaws in Bohr's interpretation of quantum mechanics and Bohr would defend his stance. Einstein would present Bohr with a thought experiment and Bohr would find a flaw in Einstein's argument, usually within a few days.

In 1948, Bohr summarized the discussions between the two men. He concluded:

'Whether our meetings have been of short or long duration, they have always left a deep and lasting impression on my mind.'

A box full of light

One of their most celebrated contests was when Einstein asked Bohr to imagine a box full of light. The box had a number

Niels Bohr and Albert Einstein

of clocks and scales set up inside it and these could be used to determine both the energy and the time of release of a single photon. Firstly the box had to be weighed, then a single photon would be released through a shutter operated by a clockwork mechanism inside the box. The box could then be weighed again and, knowing the change in mass, Einstein could calculate the energy of the photon using $E = mc^2$. He would therefore know the change in energy as well as the precise time the photon had been emitted, thus evading the uncertainty principle.

Bohr apparently spent a sleepless night trying to think of a riposte to Einstein's light box. Then the answer came to him. The photon, Bohr realized, would have a recoil when

Einstein's box contained electromagnetic radiation and a clock that controlled the opening of a shutter.

it was fired into the box, thus causing uncertainty about the position of the clock in Earth's gravitational field. Since Einstein himself had, in the general theory of relativity, shown that clocks run slower in a gravitational field, there would be uncertainty in the time recorded by the clock. Einstein had caught himself out by forgetting about his own theory!

Spooky action at a distance

In 1935, Einstein, in collaboration with colleagues Boris Podolsky and Nathan Rosen, introduced another thought experiment which argued that quantum mechanics was not a complete physical theory. Known today as the 'EPR paradox' after the three collaborators, the thought experiment was meant to tackle a peculiar feature of quantum mechanics called quantum entanglement. This said that the result of a measurement on one particle of an entangled quantum system can have an instantaneous effect on another particle, regardless of the distance between the two.

As we have seen, one of the main tenets of quantum mechanics is the idea of uncertainty – we can't measure all the features of a system simultaneously, not even in theory. We can't know, for example, position and momentum, so we have to choose to measure one or the other, not both together. Another peculiar property of quantum mechanics is called entanglement. In an example of quantum entanglement, two photons are allowed to interact so that they can subsequently be defined by a single wavefunction. (How this is accomplished need not detain us.) Once they are separated, they will still share this single wavefunction. This means that measuring one will determine the state of the other – for example, if the two quanta have a spin-zero entangled state then, if one particle is measured to be in a spin-up state, the other is instantly forced to be in a spin-down state. This is officially known as 'non-local behaviour'. Einstein called it 'spooky action at a distance'.

Quantum entanglement

Einstein accepted that quantum mechanics could accurately predict the outcomes of various experiments. He knew that it wasn't 'wrong'; but he was arguing that it wasn't yet complete and the EPR paradox was another attempt to demonstrate this – the paper was even entitled, 'Can the Quantum Mechanical Description of Physical Reality Be Regarded as Complete?' Einstein suggested there were properties of the quantum system that remained to be discovered, what he called 'hidden variables', which once known would account for the observations and explain the 'spooky action'. Bohr, naturally, disagreed with Einstein's view and passionately defended the Copenhagen interpretation of quantum mechanics.

Einstein and his co-authors began by setting out their premise, that if there was any way we could learn with absolute certainty the position of a particle, without

disturbing the particle by directly observing it, then we could
say that the particle exists in reality, independent of our
observations.

If we have two quantum entangled particles, then we
can take measurements of one particle which give us
information about the second particle without disturbing it
in any way. By measuring the momentum, say, of the first
particle, we gain precise knowledge of the momentum of the
second particle and we can do the same for other properties,
such as position.

So, the second particle, which we have not directly observed,
has properties which we know. It has a position that is real and
a momentum that is real. Since quantum mechanics tells us
we can't know both of these properties, then it appears that the
way in which quantum mechanics describes reality is, indeed,
incomplete.

The alternative, Einstein and his colleagues argued, was to
assume that the process of measuring the first particle alters

the reality of the second, instantaneously making it conform to the reality of the first particle, even if they were separated by light years of space. 'No reasonable definition of reality could be expected to permit this,' they asserted.

Wolfgang Pauli made his feelings very plain in a letter to Werner Heisenberg:

> 'Einstein has once again expressed himself publicly on quantum mechanics (together with Podolsky and Rosen – no good company by the way). As is well known, every time that happens it is a catastrophe.'

When the EPR paper reached Niels Bohr, he knew he would have to find a riposte to Einstein. According to one of Bohr's colleagues in Copenhagen, EPR came upon them like 'a bolt out of the blue. . . . Everything else was abandoned. We had to clear up such a misunderstanding at once.' It was no easy task. It took six weeks of fretting before Bohr was finally ready to make his response.

Bohr admitted that in Einstein's paper 'there is no question of a mechanical disturbance of the system under investigation'. Until then, Bohr had asserted that the disturbance caused by making a measurement of a particle led to quantum uncertainty. Now, he backed away from this position. In various arguments at the Solvay Conferences he had often rejected Einstein's thought experiments by recourse to the uncertainty principle. Now, instead, he used the concept of complementarity. The most important aspects of a quantum experiment, he said, were the conditions under which it was made. If you chose one set of conditions, for example an experiment involving wave properties, then wave properties were what you would see. If you chose something else, then you would reveal a complementary aspect to the wave properties. None of these elements, Bohr felt, were present in the EPR thought experiment and for this reason it failed to refute the Copenhagen interpretation of quantum mechanics.

If the two particles are entangled, Bohr argued, then they are effectively a single system that has a single quantum function. In addition, he noted, the EPR paper did not absolutely overrule the uncertainty principle. It was still not possible to know both the precise position and the precise momentum of the particle at the same moment. If you knew the position of A then you would know the position of B, and if you knew the momentum of A you would know the momentum of B. But what you still could not do was know these two things precisely at the same moment for A, so you couldn't know them for B either. There was no conflict with the uncertainty principle.

Einstein continued to insist he was on to something. His own theory of relativity did not permit 'spooky action at a distance'. He had forbidden it for Newton's gravity and he wasn't going to allow it for quantum mechanics. The latter, he maintained, violated two fundamental principles. The principle of separability, which maintains that two systems separated in space have an independent existence; and the principle of locality, which says that doing something to one system cannot immediately affect the second system.

Einstein's box, Schrödinger's cat

Erwin Schrödinger was among those who sided with Einstein in opposing the Copenhagen interpretation. He said of the EPR paper: 'Like a pike in a goldfish pond it has stirred everyone up.' He felt his wave equations had been misused and sometimes thought it might have been better if he had never developed them. At one point he declared of quantum mechanics, 'I don't like it, and I'm sorry I ever had anything to do with it.' In a letter to Schrödinger in 1928, Einstein complained:

> *'The Heisenberg–Bohr tranquillizing philosophy . . . provides a gentle pillow for the true believer from which he cannot very easily be aroused.'*

Einstein thought that Heisenberg's uncertainty principle might be a demonstration of the limits nature places on what we can know about a quantum object. However, he felt these limits should not be taken to imply that there wasn't a deeper, more deterministic reality, only that we had been denied access to it.

In 1935, Einstein shared a thought experiment with Schrödinger which illustrated why he felt so uncomfortable with wavefunctions and probabilities. Imagine two boxes, he said; one contains a ball and the other is empty. Before we look in one of the boxes, there is a 50 per cent chance of finding the ball. After we look, the chance of it being there is either 100 per cent or 0 per cent. But, in reality, the ball was always 100 per cent in one of the boxes. Einstein wrote:

> *'The probability is ½ that the ball is in the first box. Is that a complete description? NO: A complete description is that the ball is (or is not) in the first box. . . . YES: Before I open them, the ball is by no means in one of the two boxes. Being in a definite box comes about only when I lift the covers.'*

Clearly Einstein favoured the first answer and not the quantum mechanical second answer. Niels Bohr and the Copenhagen interpretation would say that the ball exists in a state of superposition, that it occupies both boxes until you look and see which one it's in. The act of observation makes the choice.

Einstein's answer is based on common sense, but as he had demonstrated himself with his relativity theories, common sense is not always a reliable guide to the workings of nature.

Schrödinger came up with a thought experiment of his own, which would pass into quantum folklore. It examined a core concept in quantum physics; this was that the timing of the emission of a neutron from a decaying nucleus is indeterminate until it is observed. In the quantum world, the nucleus exists simultaneously in both its decayed and undecayed states until observation collapses its wavefunction and it becomes either one or the other. It is a state of affairs that we might reluctantly accept as being true in the strange realm of the quantum, but how can these odd goings-on be scaled up into the 'real' world?

In his thought experiment, Schrödinger posed the following question: when does the system switch from its state of superposition into one definite reality?

Enter the cat.

'A cat is penned up in a box,' Schrödinger wrote, 'along with the following device: in a Geiger counter there is a tiny bit of radioactive substance, so small, that perhaps in the course of the hour one of the atoms decays, but also, with equal probability, perhaps none: if it happens . . . a relay releases a hammer which shatters a small flask of hydrocyanic acid.'

Schrödinger explained that the wavefunction of the entire system would express the situation by having in it the living or dead cat 'mixed or smeared out'. Einstein and Schrödinger were happy that their thought experiments had demonstrated their point – that there was something distinctly not right about the Copenhagen interpretation. Einstein said that a wavefunction which 'contains the living as well as the dead cat just cannot be taken as a description of a real state of affairs'. In 1948, he wrote to Max Born:

'You believe in a dice-playing God and I in perfect laws in the world of things existing as real objects, which I try to grasp in a wildly speculative way.'

For Niels Bohr, there was no reason why the rules of classical physics, which determine what goes on in the everyday world around us, should also apply to the quantum realm. What the quantum physicists were discovering was just the way things were, whether Einstein liked it or not. At some point, Bohr apparently said in exasperation to Einstein: 'Stop telling God what to do!'

Max Born

Born, expressing a disappointment felt by many physicists, said of Einstein that he was

> 'a pioneer in the struggle of conquering the wilderness of quantum phenomena. Yet later, when out of his own work a synthesis of statistical and quantum principles emerged which seemed acceptable to almost all physicists he kept himself aloof and skeptical. Many of us regard this as a tragedy – for him, as he gropes his way in loneliness, and for us, who miss our leader and standard bearer.'

Einstein never accepted the probabilities and uncertainties of quantum mechanics and sought throughout his life to find an underlying order. Nevertheless, quantum mechanics has stood up to experiment in the years since Einstein's death, and all the indicators are that he was wrong. As Stephen Hawking commented in a 1997 lecture:

> 'Einstein was confused, not the quantum theory.'

CHAPTER 22

Was Einstein the 'father of the atomic bomb'?

How big a role did Einstein play in the development of the atomic bomb?

When we think of Einstein's famous equation $E = mc^2$, we invariably connect it with the invention of the atomic bomb. The first mention of Einstein as 'father of the bomb' was probably in an article in *Time*. In case there was any doubt, on 1 July 1946 the cover of the magazine also featured an illustration of Einstein against the background of a mushroom cloud overprinted with '$E = mc^2$'.

Discovering the atom

At about the same time that Einstein was working on general relativity, Ernest Rutherford was exploring the structure of the atom at the Cavendish Laboratory in Cambridge, England. In 1907, Rutherford devised an experiment to show that the atom had a centre, which he called the nucleus. This was just two years after Einstein's paper on Brownian motion had confirmed the existence of atoms. By 1919, the year Arthur Eddington's eclipse observations confirmed general relativity, Rutherford had succeeded in transforming atomic nitrogen into hydrogen or, as the papers put it, 'split the atom'.

Ernest Rutherford

One of Rutherford's students was Niels Bohr, who formulated the model of atomic structure that explained the release of photons of different energies. This dovetailed with Einstein's idea that light was a stream of particles.

Scientists began to find evidence that Einstein's $E = mc^2$ equation was correct. The British chemist and physicist Francis Ashton, a fellow researcher at the Cavendish, made careful measurements of the atomic weights of the elements and was surprised to discover that there was a tiny amount of mass missing. This, he believed, was the energy that held the atoms together, which he called binding energy. He calculated that if it were possible to transform hydrogen, the lightest element, into helium, the next lightest 1 per cent of the mass would be annihilated and released as energy. According to Einstein's formula, there would be enough energy in a glass of water to power a steamship across the Atlantic and back again.

This was the first time that Einstein's equation had been linked with atomic research. But how could scientists gain access to that vast, untapped store of energy? Many thought it unlikely, including Rutherford, who in a speech in 1933 dismissed the idea as 'merest moonshine'. As it turned out, the attempts to unlock the energy of the atom began to show promise just as the world stood on the brink of war.

A replica of an early mass spectrometer, used for measuring the mass-to-charge ratio of one or more molecules present in a sample.

The journey to the bomb

A year before he made his 'moonshine' comment, a crucial breakthrough had been made in Cavendish's laboratory – the discovery of the neutron in the atomic nucleus. Neutrons are highly penetrative particles; if atomic nuclei were to be shattered to release their energy, neutrons would be the sparks that lit the fire.

By 1934, Irene Curie, daughter of Marie, had successfully created a new radioactive element. In Rome, in the same year, Enrico Fermi demonstrated that the slowing down of neutrons could turn them into more effective atom smashers. In 1938, Otto Hahn, working in Berlin, was baffled when he bombarded uranium with neutrons and found he was left with barium. In collaboration with Austrian physicist Lise Meitner and her nephew Otto Frisch, Hahn realized that in splitting the uranium atom he had released some of its binding energy. This was the first successful demonstration of nuclear fission.

News soon spread through the physics community. One scientist who heard the news, Leo Szilard, a Hungarian living and working in New York, was a friend of Einstein. With Einstein's help, Szilard had obtained his doctorate in physics in Berlin. He was a theoretician and, since 1933, had been researching the possibility that an atom split by a neutron might release two more neutrons, thereby triggering a chain reaction. Together with Fermi, Szilard showed this to be the case in March 1939.

In Princeton, Niels Bohr had determined it was the U-235 isotope of uranium that was easiest to fission when bombarded with neutrons. But this made up less than 1 per cent of natural uranium and it would be very difficult to separate it out. But if it could be done, Bohr warned, it would be possible to set off a devastating atomic explosion.

In Germany, meanwhile, physicists were pursuing the same ideas as Szilard. Anxious about the implications of this, Szilard pushed for a security embargo on the reporting of all nuclear research in the United States, Britain, France and Denmark. He was right to be worried because the Nazis embarked on a fission research programme in April 1939.

When word of the discovery of nuclear fission reached American theoretical physicist Robert Oppenheimer, he declared it to be 'impossible'. But after it was demonstrated to him that the experiment did work, he too immediately began investigating chain reactions. Within days, he had drawn up a rough plan for an atomic bomb. Oppenheimer would go on to mastermind the Manhattan Project, the code name for the top-secret development of the atomic bomb by the Allied powers in World War II.

Einstein and Roosevelt

On 2 August 1939, Szilard visited Einstein and urged him to write to President Roosevelt and impress on him the need for work to begin on the development of atomic weapons. This visit led to the letter signed by Einstein that was sent to Roosevelt on 11 October 1939. In it Einstein warned of the possibility of achieving a chain reaction in uranium in the near future, generating vast amounts of power. He wrote:

'This new phenomenon would also lead to the construction of bombs, and it is conceivable – though much less certain – that extremely powerful bombs of this type may thus be constructed. A single bomb of this type, carried by boat and exploded in a port, might very well destroy the whole port together with some of the surrounding territory. However, such bombs might very well prove too heavy for transportation by air.'

EINSTEIN'S REFRIGERATOR

In 1930, Einstein and Szilard set themselves the task of inventing a noiseless household refrigerator. Part of their creation was the so-called Einstein–Szilard pump, later described by Einstein as a device which used an alternating electric current to generate a magnetic guide field that moves a liquid mixture of sodium and potassium. The mixture, according to Einstein, 'moves in alternating directions inside a casing and acts as the piston of a pump; the refrigerant [inside the casing] is thus mechanically liquified, and cold is generated by its re-evaporation.' The Einstein–Szilard refrigerator was never marketed commercially, partly because there was a danger of the poisonous refrigerant leaking out.

Einstein went on to suggest that Nazi Germany might already be carrying out such research.

President Roosevelt replied that he had found the data 'of such import that I have convened a Board . . . to thoroughly investigate the possibilities of your suggestion regarding the element of uranium.'

An alpha calutron tank – a mass spectrometer used for separating the isotopes of uranium.

According to Einstein's biographer Abraham Pais, opinions are divided about the amount of influence Einstein's letter had on Roosevelt. Pais believed it was limited, pointing out that even though Roosevelt appointed an advisory committee, he did not give the go-ahead for full-scale atomic weapons development until October 1941. And it was not until 6 December 1941, the day before the Japanese attack on Pearl Harbor, that the Manhattan Project was launched.

Even if he had played a part in setting it in motion, Einstein never worked directly on the Manhattan Project. He was not invited to join it nor even officially told that it existed. J. Edgar

Hoover, the director of the FBI, was suspicious of Einstein's pacifism and believed him to be a security risk. Among other things, Hoover asserted that Einstein had supported an anti-war congress in 1932 and was pro-Soviet. In fact, Einstein had refused to attend the congress and had denounced the Soviet Union for 'its complete suppression of the individual and of freedom of speech'.

Einstein did play a small role in the Manhattan Project, however. He was asked by Vannevar Bush, one of the project's chief scientists, to help with a problem involving the separation of isotopes. After working for two days on a process in which uranium was converted into a gas and forced through filters, Einstein sent in his report. The scientists who looked at it were keen for Einstein to be given a bigger role in the project, but Bush refused: 'I wish very much that I could place the whole thing before him and take him fully into confidence,' he wrote, 'but this is utterly impossible in view of the attitude of people here in Washington.'

In December 1944, Einstein was visited by his friend Otto Stern, who had been working on the Manhattan Project. When Stern told him that the project was nearing completion, Einstein decided to write to Niels Bohr about his concern for the control of atomic weapons in the future. 'The politicians do not appreciate the possibilities and consequently do not know the extent of the menace,' he wrote. Bohr visited Einstein and urged him to be careful about making his views more widely known, warning of 'the most deplorable consequences' if information about the development of the bomb were to be made public. Einstein agreed.

On 6 August 1945, an atomic bomb obliterated the Japanese city of Hiroshima. Einstein was staying at a rented cottage in the Adirondacks when he heard the news. A few days later, following the dropping of a second bomb on Nagasaki, a report was issued detailing the development of the bomb. Much to Einstein's dismay, it laid great emphasis on his letter to Roosevelt. This is one of the reasons why popular imagination linked Einstein to the bomb, even though he had taken little part in its development.

Hiroshima following the US atomic bomb explosion

EINSTEIN'S AUCTION

To help the war effort, Einstein submitted a copy of his paper on special relativity for auction. It wasn't the original – Einstein had thrown that away after it was published. To recreate the manuscript, he had his assistant Helen Dukas read it to him aloud while he wrote down the words. The manuscript, along with one other, sold for $11.5 million.

After the war

In an interview with *Newsweek* magazine, Einstein declared:

'Had I known that the Germans would not succeed in producing an atomic bomb I never would have lifted a finger.'

In December 1945, he told an audience:

> 'The first atomic bomb destroyed more than the city of Hiroshima. It also exploded our inherited, outdated political ideas.'

Einstein was chairman of the Emergency Committee of Atomic Scientists, a group that met between 1946 and 1949. In the group's charter, he stated his belief in the need

> 'to advance the use of atomic energy in ways beneficial to mankind [and] to diffuse knowledge and information about atomic energy . . . in order that an informed citizenry may intelligently determine and shape its action to serve its own and mankind's best interest.'

CHAPTER 23

Can we find a theory of everything?

Today, theorists are still looking for a way to unite relativity and quantum mechanics. Are they close to finding an answer?

Einstein's relativity theories provide a framework for understanding the universe on the scale of stars and galaxies. Quantum theory describes how the universe works on the scale of atoms and atomic particles. Both theories have been tested to unimaginable levels of accuracy, and they appear to work, but the problem remains that we have yet to find a way to unite them. If quantum mechanics is used with general relativity to calculate the probability of a process involving gravity taking place, the answer that comes up is an infinite probability. Any answer above 100 per cent probability is meaningless. It just demonstrates that combining general relativity and quantum mechanics simply doesn't work.

Einstein spent the last 30 years of his life trying to find a way to unite electromagnetism and gravity, but he never succeeded. He was convinced there had to be a single theory that would encompass all of the universe's physical phenomena. In his Nobel Prize acceptance lecture, he said:

> 'The intellect seeking after an integrated theory cannot rest content with the assumption that there exist two distinct fields totally independent of each other by their nature.'

When Einstein began his work on a unified field theory in the 1920s, electromagnetism and gravity were the only known forces, and the only subatomic particles that had been discovered were the electron and the proton. Today, physicists know that there are two other fundamental forces; a strong force that binds together atomic nuclei and a weak force that governs radioactive decay. There is also a whole zoo of particles such as quarks, muons, gluons and neutrinos.

In Einstein's day, most physicists seemed unconcerned about formulating a theory that would unite electromagnetism and gravity. Instead, the focus was on the strange and wonderful new world of quantum mechanics which had just been opened up for exploration.

But Einstein wasn't completely alone in his quest; several other scientists turned their attention to the problem of unification. In 1918, the German mathematician Hermann Weyl had proposed a unification scheme based on a generalization of the geometry of curved space that Einstein had used in developing his general theory of relativity. Inspired by Weyl's work, another German mathematician, Theodor Kaluza, showed that if spacetime were extended to five dimensions, then four of them would encompass Einstein's general relativity equations, while the fifth dimension would be the equivalent of Maxwell's equations for electromagnetism. Oskar Klein later determined that the fifth dimension would be curled up so small that we wouldn't be able to detect it.

Einstein liked the idea of the five-dimensional approach. In 1919, he wrote to Kaluza:

'The idea of achieving unification by means of a five-dimensional cylinder world would never have dawned on me. . . . At first glance I like your idea enormously.'

Einstein also explored the approach of extending general relativity to include electromagnetism, but still keeping the four-dimensional geometry of spacetime. He persisted with both ideas for the last 30 years of his life, but never found the answers he was looking for. 'Most of my intellectual offspring end up very young in the graveyard of disappointed hopes,' he lamented in 1938.

Einstein continued to refine his ideas on a unified theory while also trying to solve what he saw as problems in his general theory of relativity, such as its prediction of black holes. One unfortunate consequence of his search for unification was that to some extent it isolated him from the rest of the physics community. There were many physicists who thought Einstein contributed nothing of importance to the field in the last 20 years of his life.

In particular, Einstein's antipathy towards quantum mechanics may have cut him off from some promising lines of research. In large part, his conviction that quantum mechanics was flawed spurred him on in his unification quest. He believed beyond doubt in a universe that exists independently and doesn't have to be observed to take form, as quantum mechanics would have it. 'Do you really believe that the Moon is not there unless we are looking at it?' he asked.

Einstein was aware of this shortcoming in his dealings with quantum mechanics, commenting in 1954 that

'I must seem like an ostrich who forever buries its head in the relativistic sand in order not to face the evil quanta.'

Towards the end of his life, and perhaps realizing that his quest was likely to be fruitless, he wrote:

'I have locked myself into quite hopeless scientific problems, the more so since, as an elderly man, I have remained estranged from the society here.'

Perhaps Einstein was simply ahead of his time. Decades after his death the search for a 'theory of everything' has become a holy grail for many physicists.

The Standard Model
During the 1960s and 1970s, particle physicists discovered that there are two basic building blocks of matter – the

fundamental particles known as quarks and leptons. Quarks are always found within larger particles, such as protons and neutrons. They are never found separately in nature and are always bound with other quarks, held together by the short-range strong nuclear force. The leptons, which include the electron, are not affected by the strong force. However, both quarks and leptons are affected by the weak nuclear force which is responsible for certain types of radioactivity. The

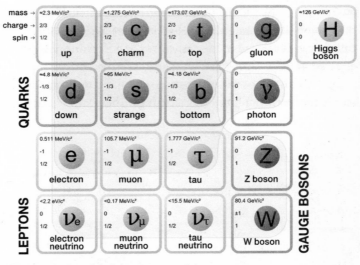

The standard model of elementary particles

strong force is stronger than the electromagnetic force over distances smaller than an atom. Gravity is the weakest of the four fundamental forces, but it acts over infinite distances. The electromagnetic force is much stronger than gravity and has an infinite range.

In 1968, Sheldon Glashow, Steven Weinberg and Abdus Salam announced a unified theory of electromagnetism and the weak nuclear force. Their electroweak theory, as it was called, suggested that the weak force was carried by particles called W and Z bosons. These were subsequently discovered in the 1980s.

Today, physicists believe that in the early universe, shortly after the Big Bang, the strong and weak nuclear and electromagnetic forces were unified. All three of these fundamental forces result from the exchange of force-carrier particles like the W and Z bosons. Each fundamental force has its own corresponding boson – the strong force is carried by the gluon, and the electromagnetic force is carried by the photon. The Standard Model of particle physics, as it is called, was developed in the early 1970s and explains how the electromagnetic, strong and weak forces, along with all their associated carrier particles, act on all of the matter particles. It works very well, but it has shortcomings. There still seems no

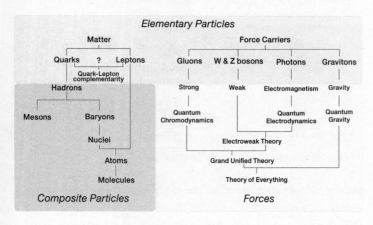

way to merge gravity with quantum mechanics and gravity is not part of the Standard Model. On the scale at which particle physics operates, the effect of gravity is so weak as to be negligible, which means that its exclusion has no effect on the predictions of the Standard Model.

String theory

Currently, one of the most hopeful candidates for a theory of everything is string theory. Not only does it promise a theory of gravity on the microscopic scale, it also seeks to provide a unified and consistent description of the fundamental structure of the universe, uniting all four fundamental forces and the fundamental particles of the Standard Model.

In December 1984, John Schwarz, of the California Institute of Technology in Pasadena, and Michael Green, of Queen Mary College, London University, published a paper showing that string theory could build a bridge across the mathematical chasm that separates general relativity and quantum mechanics.

At the heart of string theory is the idea that all of the different fundamental particles are really just different manifestations of one basic object: a string. Since the early 20th century, nature's fundamental particles, such as electrons, quarks and neutrinos, have been portrayed as examples of as-small-as-you-can-go objects, with no internal structure. String theory challenges this. It proposes that at the heart of every particle is a tiny, vibrating string-like filament. The differences between one particle and another – their mass, charges and other properties – all depend on the vibrations of their internal strings. Like a skilled violinist performing a melody, nature manifests all the particles of the atomic realm through changes in the frequency of a one-dimensional subatomic string.

Interestingly, one of the 'notes' on the string corresponds to the graviton. The graviton is a hypothetical particle that, according to quantum physics, should carry the force of gravity from one location to another, just as the photon does for the electromagnetic force. This seemed to promise a way to get gravity and quantum mechanics working together.

So, are strings 'real'? Could we go looking for them in CERN's Large Hadron Collider, for example? Unfortunately, that isn't possible. The mathematics of string theory require strings to be about a million billion times smaller than anything that the world's most powerful particle accelerators have uncovered. Unless, as physicist Brian Greene says, we can build 'a collider the size of the galaxy', we have no hope of directly detecting strings.

WHAT IS GRAVITY?

Theorists currently believe that two descriptions of gravity are equally valid. The first is that gravity is the result of the curvature of spacetime brought about by the matter in it. The second has it that gravity is an exchange of graviton force particles. Thanks to Einstein and the general theory, physicists have a workable solution for gravity that involves the gravitational field and gravitational forces as a curvature of spacetime, but there is no current quantum theory of gravity involving gravitons that is as well worked out and proven by experiment. String theory suggests that gravitons could exist but, as yet, there is no experimental proof of this.

An additional complication of string theory is that its equations require the universe to have extra spatial dimensions to make them work. String theorists took up the idea first developed by Kaluza and Klein in the early years of the 20th century when they were trying to link Einstein's gravity with electromagnetism. Perhaps, they suggested, the universe has the three big dimensions we all know and move through, but there could also be additional tiny dimensions wound up inside the 'normal' and beyond our ability to detect.

A group of theorists suggested that because strings are so small they will vibrate not just in the 'big' dimensions, but in the tiny ones as well. Audaciously, they predicted that since it is the vibrations that determine the properties of the fundamental particles which we can detect experimentally, and the vibrations are determined by the shape of the extra dimensions, there might be a way to work back to a map of the esoteric dimensions.

Unfortunately, the number of mathematically allowable shapes for the extra dimensions appeared to run into the billions. Theorist Leonard Susskind suggested that if there was no one shape that was right, then maybe they all were. Perhaps all the shapes are the right shape within their own unique universe. Our universe would just be one of a vast, perhaps infinite assembly, each with features that are determined by the shape of their extra dimensions. 'Our' universe's hidden dimensions make possible the laws of physics which led to the existence of stars and galaxies, the chemical elements, life itself. In some other dimensional configuration, different laws would apply and the universe would be a very different, probably lifeless place.

These heady ideas mirror developments in cosmology that have suggested that the Big Bang may not have been a unique event. Instead, so the theory goes, there have been infinite bangs giving rise to an infinity of expanding universes, called the multiverse. If Susskind is correct, perhaps each multiverse has its own unique complement of compact dimensions.

Can any of this be true? In theory, yes it could, though we can never have any way of knowing for sure. But the history of science demonstrates that we shouldn't dismiss ideas out of hand simply because they go against what 'common sense' suggests we should expect. With that sort of attitude we wouldn't have embraced quantum physics or Einstein and his relativity theories.

What would Einstein have made of it all? Would he have embraced the mathematics of string theory or have found some flaw in it and rejected it as he did with quantum mechanics? Would the man who was convinced that 'God does not play dice' accept the notion of an infinity of universes, the characteristics of which were each determined by a fresh roll of the cosmic dice? We will never know, but it's likely that he would have had a unique view of it.

Writing in 1953, Einstein explained:

'Every individual . . . has to retain his way of thinking if he does not want to get lost in the maze of possibilities. However, nobody is sure of having taken the right road, me the least.'

Picture Credits

CERN/Lucas Taylor: 221
Getty Images: 166 (Ted Thai)
Hubble/ESA: 134, 136, 139, 142
NASA: 153, 156, 164
Science Photo Library: 78, 104, 119, 161, 200, 208
Shutterstock: 19, 23, 27, 28, 35, 45, 51, 53, 55, 57, 59, 69, 81, 91, 94, 99, 107, 111, 115, 123, 133, 144, 149, 150, 152, 159, 173, 189, 203, 213
Wellcome Images: 21, 31, 60